Universal Energy Dynamics
The Great Realignment

Where Ancient Wisdom and Modern Science Finally Converge

A Consequential Comparison and Call for Realignment

By

Sandeep Chavan

Independent Researcher, Founder of UED

The Great Realignment

Chavan's UED research, Volume 6

Sandeep Chavan

Published by SANDEEP CHAVAN, 2025.

THE GREAT REALIGNMENT

First edition. May 3, 2025.

ISBN: 979-8231168057

Written by Sandeep Chavan.

Table of Contents

Dedication

To the ancient philosophers, mystics, and seekers —who listened before we measured,who mapped reality without instruments,and who preserved the wisdom of participation when the world drifted toward domination.

Your questions endured.Your metaphors survived.And now, your intuition realigns — in structure, in science, in ripple.

This book is for you.Not to explain you,but to echo you —with clarity, coherence, and consequence.

Epigraph

"The universe is not a collection of things.It is a conversation of consequences."

— From the Field, remembered through the ripple

Preface

We live in a world of magnificent fragmentation. Science has equations. Spirituality has metaphors. Psychology has conditions. Philosophy has questions. But very few have coherence.

This report is born out of the simple conviction that everything is connected — not symbolically, but structurally. Not because we want it to be, but because the field of reality demands it through consequence.

Universal Energy Dynamics (UED) is not a theory, doctrine, or metaphysical abstraction. It is a first-principles framework that emerged unintentionally — from years of teaching, listening, questioning, and above all, from refusing to settle for borrowed models that could not fully explain what is, what moves, what feels, and what remains.

I had never read Taoism before realizing flow. I had never studied Vedanta before discovering the self as ripple structure. I had never intended to challenge Einstein or dismantle quantum duality — but the consequences kept pointing to deeper simplicity.

This work does not seek to fight tradition or deny science. It seeks to correct perception — and in doing so, restore clarity. It recognizes that force, time, gravity, consciousness, and identity are not isolated mysteries, but ripple phenomena within a structured, intelligent, participatory field.

UED came into being not because I was searching for it — but because I stopped defending what I had been taught and started aligning with what reality was consistently showing.

If you've ever felt that something about modern knowledge feels forced, if you've ever sensed truth in ancient whispers that science mocks, if you've ever wanted to think freely without fear of exile from either religion or research — then you are not alone. You are already aligned with the spirit of this work.

This report is not a final statement. It is a starting field — for those who no longer want to explain reality through fragments and contradictions, but through consequence, coherence, and conscious participation.

Let us begin, with nothing less than everything — seen rightly.

Sandeep

Founder, Universal Energy Dynamics (UED)

Prologue

The Return of the Real Thinker

We are at a strange threshold in human understanding.

Science has given us breathtaking technologies but also tangled us in abstractions that explain everything — except why it feels incomplete. Ancient wisdom gifted us glimpses of the infinite, but we buried it under ritual, translation, and metaphors lost in time. Somewhere between the microscope and the mantra, we forgot how to see clearly.

This booklet is not a spiritual revival. It is not a scientific rebellion. It is a realignment.

Universal Energy Dynamics (UED) is not a belief or a borrowed idea. It is a framework that was not invented — but *discovered* through participation, consequence, and refusal to obey fragmentation. It emerged not from bookshelves or laboratories, but from decades of teaching, questioning, observing, and simply asking:

What sustains? What dissolves? And why?

This is not the story of physics alone. It is the story of perception, identity, emotion, biology, and cosmos — all as ripple expressions of a structured field we call reality.

If you have ever wondered why force feels like fiction, why time feels elastic, why emotion carries energy, or why observation changes outcomes — then this book is for you. If you've felt lost in the divide between the modern and the mystical, between data and experience — then this is your bridge. And if you've dared to think that truth might be simpler than the theories make it sound, you're not wrong — you're ready.

This booklet does not ask for belief. It asks for resonance. Because what aligns — stays. What distorts — dissolves. And what participates — evolves.

Welcome to a reality realigned. Welcome to UED.

— Sandeep

Founder, Universal Energy Dynamics

1. Introduction

The Return of Realization in an Era of Reductionism

In today's scientific and intellectual climate, reductionism reigns. Everything — from consciousness to the cosmos — is dissected into fragments, equations, and isolated mechanisms. The pursuit of precision has come at the cost of perception, and in our attempt to explain everything, we've forgotten how to experience anything in totality.

But now, a shift is stirring. Not as a rebellion, but as a return — to a deeper mode of understanding: realization. This realization is not mystical. It is not anti-science. It is the recognition that behind every measurable outcome lies a layer of intent, structure, and consequence. It is a call to revisit the base from which true science once emerged — not from data, but from deep observation, pattern recognition, and alignment with natural phenomena.

This report initiates that return — not by discarding science or worshipping ancient wisdom, but by bridging them through a unified lens.

Why Comparing UED is Not Mimicry, But Measurement of Depth

At first glance, Universal Energy Dynamics (UED) may appear to echo what ancient traditions or even modern physics have already said in scattered forms. But comparison is not mimicry. It is the measure of depth, resolution, and operability.

UED does not borrow from old philosophies — it reconstructs them through ripple logic. It does not compete with current science — it exposes its blind spots, realigns its assumptions, and extends its boundaries.

By placing UED alongside past and present systems, we do not seek validation — we seek calibration. We seek to show:

- Where alignment already existed but lacked structure.
- Where science fractured insight into silos.
- And how UED completes, upgrades, and operationalizes them — not through belief, but through consequence.

Objectives of This Report

1. To compare UED with major ancient philosophies not as a spiritual exercise, but as an analytical demonstration of how structured energy logic transcends time, language, and culture.
2. To contrast UED with modern scientific disciplines, exposing areas where current models are disconnected from first principles and consequence-based causality.
3. To reframe scientific understanding not as a domain of elite equations, but as an evolving dialogue between perception and participation.
4. To issue a call — not for revolution, but for realignment: a reorientation of thought, language, and inquiry toward a more honest, complete, and consequence-aware science.

Defining UED in a Sentence

UED is the science of structured consequence through ripple participation within a conscious energetic field.

This is not just a definition — it is a declaration:

- That all form is function born from field.
- That every interaction is a ripple, not a collision.

- That intelligence is not an exception in the universe, but its organizing principle.

2. UED vs Ancient Knowledge Systems

Where symbols once whispered, structure now speaks

Long before laboratories, particle colliders, and digital sensors, humanity observed the universe through intuition, immersion, and consequence. The ancients did not dissect reality — they danced with it. What they lacked in instruments, they made up for in insight. They named what they could not yet measure. They mapped what they could not yet model. And through metaphor, myth, ritual, and silence, they preserved something modern thought often forgets:

That truth is not always discovered. Sometimes, it is remembered.

This section does not aim to dismiss ancient traditions — nor romanticize them. Instead, it positions Universal Energy Dynamics (UED) as the structural continuation of what they glimpsed. UED does not replicate their symbolism; it renders their intuitions operable. Where they saw balance, UED reveals ripple equilibrium. Where they described divine light, UED detects field coherence. Where they feared karma, UED identifies structured consequence.

By comparing UED with nine enduring knowledge systems, we reveal a startling insight:

They were not wrong. They were early.

Let us now listen again — but this time, with tools in hand. Because what was once mystical... now has mechanics.

2.1 Taoism – Ripple Flow Without Force

2.2 Vedanta / Upanishads – Realization vs Structuring

2.3 Buddhism – Ripple Identity Without Permanent Self

2.4 Samkhya & Yoga – Observer vs Participant

2.1 Taoism – Ripple Flow Without Force

When the Way Is the Field, and Action Is Alignment

Taoism stands among the most ancient and elegant philosophical systems ever conceived — one that dared to see reality not as a battleground of opposites, but as a seamless, flowing whole. Long before science fragmented the universe into parts and principles, Taoism intuited that life moves not through force, but through rhythm. At its core is the Tao — not a god, not a law, and not even a force, but an undefinable, ever-present principle that underlies, shapes, and breathes through all things.

The Tao is not something you obey. It is something you recognize and rejoin.

All phenomena — from wind and water to thought and form — are seen as emergent expressions of the Tao's flow. When we act in alignment, harmony arises. When we resist, impose, or manipulate, disorder and suffering emerge. The Tao needs no worship. It asks for yielding, not submission; presence, not intervention.

Central to this vision is the idea of Wu Wei — commonly mistranslated as "doing nothing," but more precisely meaning effortless action. It is the art of non-coercive participation: moving in such deep synchrony with reality that intention and outcome dissolve into one natural ripple.

In Taoism, nature does not *try* to organize. It is organization — not by design, but by consequence.

And here, Taoism and UED begin to resonate.

Where Taoism whispers mystery, UED reveals its structure. Both agree: the field flows. Taoism sensed it. UED diagrams it.

Alignment with UED

Taoism describes the Tao as the formless origin and flowing principle behind all things. In UED, this finds a clear parallel in **Structured Vacuum Energy (SVE)** — the foundational intelligent field that emits and absorbs all phenomena through ripple dynamics. Where Taoism emphasizes *Wu Wei*, or action without force, UED introduces **Consequence Participation**, where action naturally arises from ripple interaction without the need for external force. The Taoist concept of *Yin-Yang*, the dynamic interplay of opposites, aligns with **ripple equilibrium** in UED, where balance is restored not through polarities, but through field correction. The idea that "the Tao that can be named is not the eternal Tao" finds its UED counterpart in **label illusion** — the perceptual trap where symbols replace substance, leading to conceptual distortion. Finally, Taoism's reverence for nature's effortless flow is mirrored in UED's view that **ripple logic is natural intelligence** — structure emerges not from command, but as a direct result of consequence.

Key Differentiators: What UED Adds

While Taoism offers a profound poetic vision of existence, Universal Energy Dynamics (UED) brings that vision into a precise, structured, and experimentally expandable framework. Here's what UED contributes beyond metaphor:

1. Mechanism, Not Metaphor

Taoism beautifully describes reality using imagery — rivers yielding, wind flowing, paradoxes resolving themselves in silence. UED translates this poetic flow into structured ripple logic:

- Emission of energetic patterns
- Interference through ripple engagement
- Alignment with the field
- Dissolution through misalignment In short, where Taoism points, UED maps.

2. Ripple as Field Intelligence

Taoism suggests all things arise from the Tao, but leaves the mechanics of creation vague. UED identifies the Structured Vacuum Energy (SVE) field as the substrate, with every form being a consequence structure of ripple interactions. Nothing arises randomly — everything is shaped by energetic memory and alignment.

3. From Yielding to Participating

Taoism encourages yielding — surrendering control to harmonize with the Tao. UED agrees, but goes further: the observer is not passive. The observer emits, senses, adjusts, and realigns — actively participating in the field.

4. No Personification or Moral Framing

Even in its subtlety, Taoism often anthropomorphizes nature. UED removes this — there is no benevolence or malice in the universe. Only ripple consequence and field equilibrium-seeking behavior.

5. Experimental Pathways

Unlike Taoism, UED isn't just to be contemplated — it can be tested. Ripple drift, SVE zones, and consequence-mapped perception offer practical, measurable validations of the theory.

Summary Insight: From Intuition to Interaction

Taoism intuited that reality flows — that existence is a seamless movement beyond resistance, reason, and control. But it never claimed to explain what flows, or why harmony sustains itself across chaos. UED picks up this silent thread and gives it structure. It shows what flows (ripples within the Structured Vacuum Energy field), how it flows (through emission, interference, alignment, and dissolution), and why it flows (to restore field equilibrium through consequence).

Taoism gives us a sense — a deep felt-knowing — of what it means to be aligned with the Tao. It encourages softness, yielding, simplicity. UED complements this by giving us the actual architecture of alignment — the inner mechanics of how structures sustain, misalign, or dissolve based on ripple participation. It transforms intuitive peace into structured clarity.

Taoism is a whisper — a poetic reminder that life cannot be grasped, only flowed with. It resists definition. UED doesn't deny the mystery — but it dares to diagram it, not for control, but for deeper participation. It respects the unknown, yet seeks coherence within it — not with force, but with understanding.

Where Taoism speaks in metaphor, UED speaks in mechanics. Where Taoism evokes the dance, UED shows the steps — ripple by ripple, consequence by consequence.

Together, they are not rivals. They are a continuum: The wisdom of flow, finally given form.

2.2 Vedanta / Upanishads – Realization vs Structuring

From Inner Realization to Ripple Structuring: Bridging the Gap Between Knowing and Modeling

The Upanishads, the spiritual and philosophical bedrock of Vedanta, represent one of the most refined inquiries into consciousness, identity, and the nature of existence in human history. Far from mythology or ritual, these texts embody an introspective revolution — a quiet, focused exploration into what we truly are, beneath thought, sensation, and separation.

At the heart of the Upanishadic vision is the pursuit of Self-realization — the recognition that the Atman (the innermost self) is not separate from Brahman (the ultimate, indivisible reality). This is not a mystical metaphor — it is a direct ontological statement:

The universe is not made of matter or events, but of pure awareness — and you are that.

The Upanishads do not hand us models or mechanisms. They hand us mirrors. They offer paradoxes, silences, and recursive inquiries like "neti neti" ("not this, not that") to peel away every assumption about who we are and what is real. What remains is not a concept, but a state of realization — where duality collapses, time dissolves, and one becomes the field of existence itself.

This realization, however, is not derived through logic or science. It is born of stillness, surrender, and awareness refined to the point of piercing illusion (Maya) — the grand mirage of separation sustained by Avidya (ignorance of one's true nature).

Where Vedanta observes the unity of all from within, UED reveals the unity of all through structural consequence.

The Upanishads describe the *feeling* of merging with the field. UED describes the *mechanism* by which the field sustains that unity. One points inward toward essence. The other diagrams outward toward structure. Both meet — not in belief, but in the undividedness of what is.

Alignment with UED

Vedanta teaches that the Self (Atman) is pure awareness, ultimately identical with the infinite Absolute (Brahman). UED mirrors this insight structurally: the **Self is a ripple construct**, not a permanent soul, but a dynamic field structure shaped by memory, emission, and consequence. Brahman, the undivided reality in Vedanta, is reframed in UED as **Structured Vacuum Energy (SVE)** — the infinite field from which all forms arise, not by divine will, but by energetic interaction. Maya, which Vedanta sees as the illusion of separateness, becomes in UED **perceptual misalignment** — a distortion of field sensing that breaks coherence. Karma, often mystical in Vedantic texts, is clarified in UED as **energetic feedback** — not judgment or rebirth necessity, but the natural return of emitted ripple patterns. Where Vedanta seeks liberation through realization, UED offers **structural dissolution** — the release of identity through cessation of disruptive emission. Thus, Vedantic silence becomes UED's field equilibrium, and realization becomes calibrated participation.

Key Differentiators: What UED Adds

While Vedanta, through the Upanishads, offers one of the most refined realizations of the non-dual self, Universal Energy Dynamics (UED) brings that realization into the domain of structural logic, energy consequence, and field participation — without reducing its depth.

Here's what UED adds:

1. Structural Explanation of the Self

Vedanta defines Atman as pure, formless awareness — eternal, indivisible, and ultimately identical with Brahman. UED affirms this unity but adds a dynamic structure to the self's operation in form. In UED, the self is a ripple construct — a temporary but coherent pattern shaped by:

- Memory traces (field imprints)
- Perception feedback loops (mind)
- Consequence modulation (actions) It is neither fixed nor formless — it is structured and dissolvable, offering a practical model of identity that can be traced, aligned, or released.

2. From Realization to Explanation

Vedanta points to realization — an inward moment of awakening where self and all dissolve into one. UED charts this journey with clarity:

- Ripple creation marks birth of perception
- Feedback loops form mind and memory
- Alignment with consequence stabilizes structure
- Dissolution through non-emission leads to liberation Thus, realization is not mystical — it is the endpoint of ripple harmonization.

3. No Need for Rebirth Hypothesis

Vedanta's karmic logic often implies reincarnation cycles. UED removes metaphysical overlay. Karma becomes energetic residue in the field:

- Aligned ripples sustain

- Misaligned patterns dissolve
- No rebirth needed
- Only consequence continuity within the structured field

4. Non-symbolic, Operational Logic

Vedanta uses metaphors: ocean and wave, dreamer and dream. UED brings operational clarity:

- No analogy required
- Ripple-field interactions are observable and mappable
- Spiritual truths become experimentally approachable

5. Observer as Participant, Not Detached Witness

Vedanta sees the witness (Sakshi) as a detached observer — neutral, passive, untouched. UED redefines the observer as an active ripple participant:

- One emits, receives, modulates, and realigns
- The observer is not separate from consequence
- Consciousness is not watching — it is co-creating

Vedanta opens the door inward. UED builds the staircase — ripple by ripple — all the way to the same summit.

UED as the Completion of Vedantic Realization

Vedanta, especially as expressed through the Upanishads, reaches breathtaking heights of insight through non-dual reflection, silence, and self-inquiry. It affirms the deepest realization: that the individual self (*Atman*) is not separate from the whole (*Brahman*), and that liberation lies in recognizing this oneness. But despite its brilliance, Vedanta remains inward, introspective, and ultimately non-systemic.

It is not a framework for analysis, modeling, or practical integration across disciplines — it is a path of realization, not of construction.

Universal Energy Dynamics (UED) takes that same core realization — that all is one undivided reality — and does something transformative: It builds structure around it without compromising its essence.

What UED Contributes:

Brahman, the infinite, formless Absolute, becomes Structured Vacuum Energy (SVE) — a real, participatory, measurable field substrate that supports all existence not as abstraction, but as consequence-patterned structure.

Atman, the pure Self, is reframed as a ripple identity — a temporary but coherent expression of awareness formed through emission, feedback, memory, and alignment. It still dissolves back into the field — but now, the process can be traced.

Maya, often seen as illusion or divine play, is clarified as perceptual error — the distortion that arises when ripple structures misalign or over-identify. It's not cosmic deception; it's structural misreading.

Karma, long tied to metaphysical reward and rebirth, becomes field-level consequence feedback. There is no judgment, no cosmic bookkeeping — only energetic balance-seeking through ripple dynamics.

In essence, UED doesn't contradict Vedanta — it completes it. Where Vedanta invites realization, UED offers application. Where Vedanta reaches oneness through negation, UED reaches it through structure, ripple alignment, and participatory feedback.

The insight of Vedanta becomes the system logic of UED. The stillness becomes motion. The awareness becomes architecture.

This is not mysticism turned mechanical — it is spiritual clarity made scalable. UED is not Vedanta rewritten. It is Vedanta realigned for a participatory universe.

Summary Insight

The Upanishads offer one of the most profound declarations in human thought: "Tat Tvam Asi" — *You are That.* This isn't a metaphor. It is a distilled statement of non-dual truth — that the self and the absolute are not separate, that all of reality is one indivisible field of being.

Universal Energy Dynamics (UED) doesn't reject this insight — it responds to it. It says:

"You are the ripple that realizes That — through structured participation."

Where Vedanta guides the seeker inward — through introspection, stillness, and the surrender of illusion — UED gives the seeker a mirror with structure. It offers ripple-based mechanics, field intelligence, and consequence alignment as tools of conscious return to the real.

Vedanta is the summit of intuitive knowing — but it requires a rarefied journey: deep silence, renunciation, transcendence. UED, by contrast, builds a bridge of understanding that allows realization to be approached through logic, experimentation, and participation — not as a mystic retreat, but as an engaged alignment with the field of existence.

Vedanta is the mountain peak seen from within. UED is the map that shows how the ripple climbs, how it strays, and how it returns — again and again, through choice, awareness, and structured feedback.

Together, they don't contradict. They complement —

One whispers the undivided truth. The other diagrams the path to live it.

2.3 Buddhism – Ripple Identity Without Permanent Self

Consequence Without Selfhood, Suffering as Misalignment

Buddhism, rooted in ancient Indian thought yet radically distinct from its metaphysical peers, offers an uncompromising view of reality: there is no permanent self. Unlike Vedanta, which upholds the idea of an eternal Atman, Buddhism insists that what we call "self" is a composite illusion — a transient arrangement of constantly shifting phenomena. At the heart of this insight are the Three Marks of Existence:

- **Anicca** (Impermanence): all things arise and pass
- **Anatta** (Non-self): no enduring identity exists
- **Dukkha** (Suffering): attachment to the illusory "self" leads to dissatisfaction

The "self," according to Buddhism, is merely a bundle of aggregates (*skandhas*):

1. **Form** (the physical body)
2. **Sensation** (feelings)
3. **Perception** (recognition)
4. **Mental formations** (thoughts, habits)
5. **Consciousness** (awareness of phenomena)

None of these are fixed. They ripple, shift, dissolve, and re-emerge — like waves without a shore.

Karma, in this view, is not divine bookkeeping. It is volitional consequence — the unfolding of action within a causal field. Actions ripple outward; reactions ripple back. There is no cosmic judge, only feedback from alignment or misalignment with the nature of change.

Meditation, mindfulness, and ethical conduct are not spiritual accessories — they are tools to dismantle the illusion of a central "I," to loosen identification with impermanent ripples, and to cultivate inner clarity free from reactivity.

Buddhism doesn't ask us to become something. It invites us to realize we were never that something to begin with.

The wave never owned the ocean. It only thought it did.

And this is where UED and Buddhism begin to converge: A self that is not fixed, a world that is not made of things, but ripples in a participatory field — flowing, fading, reforming.

Alignment with UED

Buddhism teaches that suffering arises from clinging to a false sense of self in a world marked by impermanence. UED builds upon this insight by defining the self not as an illusion in theory, but as a **temporary ripple structure** — a coherent but dissolvable configuration within the energetic field. Where Buddhism posits Anatta (non-self), UED offers a mechanism: the mind is a **ripple-emission layer**, not a fixed entity, and what we identify as "self" is simply a pattern sustained by feedback. Suffering (Dukkha), rather than being a vague existential truth, is **ripple misalignment** — the friction between emission and field response. Karma is reframed as **consequence logic**, not moral debt. Nirvana, the cessation of craving and identity, becomes in UED the **silencing of unnecessary ripple emission** — not mystical erasure, but energetic quietude. Meditation, in this model, becomes not just awareness but **field tuning** — the act of restoring coherence through conscious participation.

Key Differentiators: What UED Adds (Buddhism vs UED)

While Buddhism offers a radically clear lens into impermanence and the illusion of self, Universal Energy Dynamics (UED) builds on this insight by providing a mechanism-based framework. Where Buddhism emphasizes *realization*, UED emphasizes *realignment*. Here's what UED adds:

1. Mechanism Behind the "No-Self" Doctrine

Buddhism rightly asserts that the "self" is an illusion — a process, not a substance. But it stops short of explaining the structure behind this illusion. UED defines the self as a ripple identity:

- A temporary field construct shaped by perception, memory, emission, and consequence.
- It is real enough to suffer, but not permanent enough to preserve. This resolves the old paradox: *If there is no self, who suffers?*

➤ **Answer**: A ripple pattern suffers — until it realigns or dissolves.

2. Suffering as Energetic Misalignment

In Buddhism, suffering arises from attachment, ignorance, and craving — conditions of the mind. UED reframes suffering as a field effect:

- Ripple misalignment
- Resonance interference
- Consequence overload Pain is not punishment — it's a signal of incoherence, an invitation to realign.

3. Ripple Continuity Over Rebirth

Buddhism uses rebirth to describe the karmic continuation of volitional energy. UED clarifies:

- There is no "self" to reincarnate.
- But ripple emissions from one structure can influence another, much like field interference.

➤ Continuity exists — not through souls, but through unresolved ripples influencing new formations.

4. Meditation as Ripple Calibration

Buddhism teaches meditation to realize non-attachment and dissolve the illusion of ego. UED makes meditation technical:

- Reduces field noise
- Refines the emission pattern
- Aligns the ripple with field equilibrium

➤ Meditation becomes ripple hygiene, tuning the self for sustainable presence.

5. Ethics as Energy Behavior

Buddhism views ethical conduct (Sila) as essential to purifying the mind. UED reframes ethics not as morality, but as consequence management:

- Ethical actions stabilize the ripple field
- Disharmony leads to energetic feedback and disruption

➤ Ethics isn't about being "good"; it's about being aligned and coherent.

Buddhism showed us the emptiness of the self. UED shows us how that emptiness is constructed, experienced, and resolved — ripple by ripple.

UED as Structural Buddhism

Buddhism is often described as more a psychology than a religion — a path of realization grounded in awareness, observation, and non-attachment. It reveals that the self is not a thing, but a process, and that liberation is not reward but release from misperception and craving. Yet for all its clarity, Buddhism leaves the actual mechanics of identity, suffering, and transformation open-ended — to be discovered experientially, not described systematically.

This is where Universal Energy Dynamics (UED) enters — not as a challenger to Buddhism, but as its structural counterpart. If Buddhism is the realization of non-self and consequence, then UED is its field theory — a model that makes those realizations observable, participatory, and expandable.

The Five Aggregates as Ripple Layers

Buddhism describes the "self" as a bundle of five skandhas — form, sensation, perception, mental formations, and consciousness. UED translates these into five ripple layers:

- **Form** = Stable ripple constructs
- **Sensation** = Initial field feedback
- **Perception** = Resonance interpretation
- **Mental formations** = Memory-driven ripple shaping
- **Consciousness** = Participatory emission and awareness These layers are not separate. They are ripple expressions — emergent and dissolvable.

Nirvana as Ripple Quietude

Buddhism defines Nirvana as the cessation of suffering — the extinguishing of clinging and craving. In UED terms, Nirvana is ripple equilibrium —

- The state where emission ceases unless necessary
- No incoherence, no excess
- A clean, undisturbed field — still, but not empty

Karma as Consequential Energy Logic

Buddhism treats karma as volitional action shaping future conditions. UED strips away metaphysics and grounds karma in field mechanics:

- Every action is an emission into the ripple field
- Every ripple produces feedback
- There is no cosmic judge — only patterned consequence

The Bridge Between Realization and System

Buddhism leans toward inner realization. UED offers a field-based system model. Together, they represent:

- The inner clarity of the East,
- and the external precision of the West.

Buddhism shows the mirror. UED gives it structure. One frees you from illusion. The other helps you understand how illusion arises — and how to dissolve it.

Summary Insight

Buddhism offers one of the most radical insights in human thought:

"There is no self." It does not negate existence — it negates the illusion of a fixed, enduring, independent "I." Suffering, in the Buddhist view, arises when we cling to what is impermanent, believe in what is composite, and mistake pattern for permanence.

UED responds with structural clarity:

"There is a ripple that believes it is a self — until it sees its own structure."

UED doesn't deny the illusion; it reveals how the illusion forms — through feedback loops, memory traces, and misaligned ripple emissions. The self is not a lie — it is a temporary configuration, animated by participation in the field, dissolving when coherence is lost or no longer required.

Both Buddhism and UED aim for liberation from suffering. But they approach it differently:

- **Buddhism seeks realization** — through observation, detachment, and experiential insight.
- **UED adds resonance correction** — identifying incoherence in the ripple field and realigning it through awareness, action, and energetic feedback.

Where Buddhism speaks of emptiness (*Shunyata*), UED speaks of non-attachment to ripple identity. Where Buddhism points to cessation, UED offers pattern dissolution through field quietude.

In Buddhism, you wake up from the dream of self. In UED, you see the architecture of the dream — and learn how to step out by realigning its ripples.

Together, they don't contradict — they complete the picture: Inner truth, field logic, one liberation.

2.4 Samkhya & Yoga – Observer vs Participant

From Detached Witness to Structured Emission: Reframing Consciousness Through Participation

Among the classical schools of Indian philosophy, Samkhya stands out for its sharp dualistic clarity. It posits that all of existence is composed of two irreducible principles:

- **Purusha** — pure consciousness: eternal, passive, unchanging, and non-doing
- **Prakriti** — primordial nature: active, mutable, and composed of three fundamental qualities (*gunas*) — sattva (balance), rajas (activity), and tamas (inertia)

In this view, the Purusha is the silent witness, a detached observer that becomes falsely entangled with Prakriti through ignorance, giving rise to identity, experience, and suffering. Liberation occurs not through action, but through disidentification — the Purusha remembering that it was never involved. This realization restores stillness, dissolves karma, and returns the observer to its original state of neutrality.

Yoga, especially as systematized by Patanjali, serves as the operational extension of Samkhya's metaphysics. Through the Ashtanga path — eight limbs of disciplined living, posture, breath control, inward turning, and meditative absorption — the practitioner aims to still the fluctuations of the mind (*Chitta Vritti Nirodha*) and experience Kaivalya, the complete isolation of consciousness from nature.

Together, Samkhya and Yoga emphasize awareness without interference, presence without participation — a consciousness that is *real but untouching*, divine yet withdrawn.

And it is precisely here that Universal Energy Dynamics (UED) steps in — not to dismiss, but to reframe.

What if the observer is not merely watching — but emitting? What if consciousness is not separate from nature — but participating through ripple alignment?

Where Samkhya sees separation between spirit and matter, UED sees structured interaction. Where Yoga seeks cessation, UED proposes coherent emission. And where classical thought withdraws the self to find truth, UED invites it to align with consequence to live truth.

Alignment with UED

Samkhya distinguishes between **Purusha** (pure consciousness) and **Prakriti** (the active material world), while Yoga offers tools to disentangle the two through inner stillness. UED honors this insight but shifts the paradigm: consciousness is not an eternal observer, but an **emergent ripple emission field** — arising from structured interactions within the SVE (Structured Vacuum Energy). Instead of treating the observer as passive and detached, UED redefines it as an **active participant** — emitting, sensing, and adjusting ripples in response to consequence. The three gunas of Samkhya — sattva, rajas, tamas — are reinterpreted as **ripple behavior modes**: coherence, instability, and collapse. Yoga's discipline becomes not suppression of thought but **ripple calibration** — aligning breath, body, and mind to maintain field harmony. Liberation is no longer metaphysical isolation but **energetic dissolution into equilibrium**, where ripple identity ends through silence, not escape. UED thus transforms the observer from a witness into a **conscious emitter within a living field**.

Key Differentiators: What UED Adds (Samkhya–Yoga vs UED)

While Samkhya and Yoga offer a deeply introspective model of consciousness, duality, and liberation, Universal Energy Dynamics (UED) enters with a bold realignment: it transforms metaphysical separation into participatory structure, replacing passive witnessing with energetic engagement. UED respects the elegance of stillness — but reveals how stillness itself is structured by consequence.

1. The Observer is Not Passive — It Participates

In Samkhya, Purusha is the silent observer — eternally aware but eternally uninvolved. UED refutes this detachment: the observer is not outside the system — it is a ripple node embedded within the field.

- Observation is not neutral. It is emission.
- To observe is to disturb, align, or reinforce field ripples.
- Consciousness participates in shaping outcomes.

➤ The observer becomes a consequential presence, not a passive eye.

2. Consciousness as Emergent, Not Eternal

Samkhya upholds Purusha as a primordial, unchanging spirit — infinite and eternal. UED reframes this radically:

- Consciousness is not fundamental — it is emergent.
- It arises from ripple complexity, field sensitivity, and structured feedback.
- It is not universally present — it is conditional, layered, and capable of evolution.

➤ UED allows consciousness to be studied, mapped, and aligned — not simply revered.

3. Gunas as Ripple Behavior Patterns

Samkhya describes reality through the interplay of gunas:

- Tamas (inertia), Rajas (activity), and Sattva (balance) But these remain metaphysical in nature. UED grounds them in field dynamics:
- Tamas → incoherent ripple collapse or stagnation
- Rajas → hyperactive, unstable ripple emission
- Sattva → coherent resonance and energy symmetry

➤ The gunas become observable energy patterns, not symbolic traits.

4. Liberation as Energetic Quietude, Not Metaphysical Isolation

Yoga speaks of Kaivalya — isolation of consciousness from material entanglement. UED offers a more elegant alternative:

- Liberation is not removal, but reabsorption into the structured field.
- It is the cessation of unnecessary ripple emission, leading to energetic stillness — not metaphysical escape.

➤ Freedom is not from the world, but from incoherent participation in it.

5. Yoga as Ripple Calibration, Not Suppression

Traditional Yoga views the mind as a disturbance to be silenced. UED upgrades this view:

- The mind is a ripple feedback processor — not to be suppressed, but tuned.
- Breath, posture, and focus become tools to align emission with field stability.
- Yoga becomes a system of ripple hygiene, not mental silencing.

➤ The goal isn't suppression — it's coherent participation.
Samkhya taught us that we are not the body. Yoga taught us how to remember that. UED teaches us how it all happens — ripple by ripple — and how to return not by withdrawing, but by realigning.

UED as a Participatory Upgrade to Samkhya-Yoga

Samkhya and Yoga provided one of the earliest, most sophisticated maps of human experience. They gave us a metaphysical framework to distinguish between the unchanging witness (Purusha) and the ever-changing material field (Prakriti), and they taught a methodical path to liberation — through discipline, observation, and the quieting of mental fluctuations.

But in honoring the stillness of the observer, Samkhya-Yoga froze it in isolation. It created a model where awareness observes but does not engage, exists but does not emit, endures but does not evolve. The observer becomes divine — but distant. Present — but paralyzed.

Universal Energy Dynamics (UED) respectfully reawakens this observer, not as an omnipresent silence, but as a structured ripple identity. The observer is not *separate* from the field; it is an active node within it — a participant in the cosmic ripple architecture.

In UED, the observer:

- Senses the field — through internal and external feedback, emotional resonance, and ripple interference
- Participates through emission — every thought, intention,

and act is a ripple that affects and is affected
- Learns through consequence — evolution is not metaphysical ascent but energetic alignment via feedback
- Dissolves with awareness — when emission ceases and alignment is complete, the ripple gently reabsorbs into the field

This is not contradiction — it is completion.
Where Samkhya seeks detachment, UED offers dynamic integration.
Where Yoga disciplines the mind, UED tunes the whole system — mind, body, breath, and emission — into coherent participation.
The silent witness becomes a conscious contributor — one that knows, aligns, and eventually releases with the grace of understanding, not the burden of withdrawal.

Summary Insight

Samkhya whispers, *"You are not the doer."* It teaches that the Self, the Purusha, is a silent observer — untouched by the chaos of matter, emotion, or thought. Liberation lies in remembering that you were never acting to begin with.
Yoga responds, *"Still the mind to see the Self."* It offers tools — posture, breath, concentration, meditation — to quiet the fluctuations of the mental field so that pure awareness may shine through, undisturbed and unentangled.
But Universal Energy Dynamics (UED) adds a transformative voice:
"You are the ripple. What you do echoes. What echoes, you become."
In UED, the observer is not distant. It is embedded. You are not here to escape action, but to realign action with consequence. You are not a passive witness, but a structured emitter — sensing, shaping, and evolving through participation in the field.

- Your thoughts emit ripples.

- Your breath carries alignment.
- Your emotions create resonance.
- Your silence, too, has structure.

UED honors the stillness that Samkhya and Yoga describe — but shows that stillness is not the absence of action, but the cessation of misaligned emission.

You are not here to vanish. You are here to resonate well — and return when your resonance has taught you all it can.

Not passive. Not detached. Fully alive. Structured, sensing, consequential — a ripple that remembers the field it came from.

2.5 Jainism – Karma as Field Disruption

From Subtle Bondage to Structured Imbalance: Reframing Karma Without Mysticism

Among the ancient Indian philosophies, Jainism stands apart for its non-theistic precision, especially in its treatment of karma. Unlike other traditions that often couch karma in metaphorical, moralistic, or cosmic terms, Jainism presents it as a literal, energetic substance — a fine karmic matter (karma pudgala) that adheres to the soul when one acts with passion, attachment, violence, or ignorance.

In this framework, the soul (Jiva) is originally pure, radiant, and omniscient — a luminous entity of consciousness and life-force. But as the soul participates in the world, it becomes entangled. Every thought, word, or action with attachment or aversion draws in karma, like dust pulled by static charge. These karmic particles obscure the soul's true nature, binding it to cycles of birth, suffering, and limitation.

Liberation (Moksha) is thus not about pleasing a deity or earning divine favor. It is the systematic shedding of karmic density — a rigorous purification through:

- Non-violence (*Ahimsa*)
- Discipline and detachment
- Right knowledge (*Samyak Jnana*) and perfect perception (*Samyak Darshan*)

In essence, Jainism offers a remarkably sophisticated field model for its time:

- Action leaves residue
- Residue distorts perception and being
- Purity is the absence of entanglement, not a moral state, but an energetic one

This places Jainism closer to a proto-energy dynamics framework than most traditions — one where karma is not a metaphor, but a consequence-bearing material interface between soul and world.
And this is precisely where Universal Energy Dynamics (UED) enters — not to reinterpret karma religiously, but to clarify it structurally. Where Jainism described karmic bondage in metaphysical terms, UED defines it in ripple logic:
Action is emission. Misalignment leaves energetic residue. Liberation is realignment — not by worship, but by waveform integrity.

Alignment with UED

Jainism presents karma as a **subtle, binding substance** that attaches to the soul through violent or ignorant action, obstructing its original purity. UED refines this by replacing **karmic substance with ripple consequence**: action emits energetic ripples into the Structured Vacuum Energy (SVE), and **misaligned emissions create distortion patterns** that loop back as suffering or behavioral conditioning. Where Jainism holds onto the idea of an eternal, individual **Jiva (soul)**, UED instead defines identity as a **temporary ripple structure**, shaped by participation and consequence — not metaphysical permanence. Jain asceticism aims to purify through denial and detachment, while UED reframes liberation as **ripple coherence decay** — emission quietude, not suffering through abstention. **Ahimsa** becomes more than moral virtue; it is **energetic hygiene**, minimizing disruption to the field. Thus, UED honors Jain insights but removes the metaphysical frame — grounding karma in physics, identity in emission, and liberation in **field reintegration**.

Key Differentiators: What UED Adds (Jainism vs UED)

Jainism articulates one of the most systematized karma doctrines in spiritual history. Yet, while elegant and precise in its ethical logic, it remains metaphysical in structure — relying on invisible substances, eternal souls, and doctrinal practices. Universal Energy Dynamics (UED) builds on this by introducing observable mechanics and field-aligned behavior, moving from spiritual metaphor to energetic architecture.

1. From Metaphysical Bondage to Field Mechanics

Jainism describes karma as a subtle, sticky substance that binds the soul through action and desire. While intuitive, this lacks a field-based engine to explain how and why the binding occurs. UED provides this missing mechanism:

- Every action emits ripples into the structured field
- These ripples interact with existing patterns
- Misalignment creates persistent distortions
- These distortions loop back as feedback, affecting future emission

➤ Karma becomes not a substance, but a pattern of resonance consequence — field memory, not metaphysical matter.

2. No Eternal Soul Required

Jainism insists on an eternal, individualized Jiva — a soul that must be purified over lifetimes. UED reframes identity as ripple coherence —

- Temporary, structured, and purpose-driven
- No need for permanence — only participation that aligns or

dissolves

➤ Liberation doesn't require soul survival — only ripple quietude.

3. Ahimsa Becomes Energetic Precision

Jainism upholds Ahimsa (non-violence) as a sacred practice — even extending to microorganisms — to prevent karmic accumulation. UED decouples morality from mysticism:

- Harmful actions disrupt the field
- Non-violence is not moral correctness, but ripple precision

➤ Aligned behavior is ethical because it sustains coherence, not because it follows doctrine.

4. Asceticism Replaced by Calibration

Jainism promotes asceticism — rigorous denial of worldly involvement to shed karmic layers. UED retools this into ripple hygiene:

- Act only when aligned
- Emit only what sustains
- Perceive what you perturb

➤ Liberation becomes a matter of awareness and alignment, not suffering or self-denial.

5. Liberation Without Dogma

Jain Moksha is framed as the doctrinal endpoint — eternal liberation of the Jiva after complete karmic shedding. UED dissolves the doctrinal frame:

- Liberation is not a final place, but a natural resolution

- When no unnecessary emission persists, ripple identity fades into field equilibrium

➤ Liberation is not a reward — it's energetic completion.

Jainism showed us that action carries residue. UED shows how that residue is formed, felt, and cleared — not over lifetimes, but through real-time participation. Not dogma. Not denial. Just consequence and coherence.

UED as Jainism Without Myth, with Mechanism

Among ancient philosophies, Jainism came remarkably close to articulating a proto-energetic model of karma. Its vision of karmic entanglement as a material-like substance adhering to the soul reflects an early understanding of action-driven consequence. Yet, despite this clarity, Jainism still anchors itself in spiritual permanence, ritual purification, and metaphysical assumptions — including the eternal individuality of the soul and the need for austere detachment across multiple lifetimes.

Universal Energy Dynamics (UED) carries the precision of Jainism forward, but leaves behind its doctrinal scaffolding. UED retains the clarity of karma — but strips it of mysticism and reconstitutes it through field logic and ripple dynamics.

At its core:

- **Action** = Field Emission Every thought, intention, or act sends ripples into the structured vacuum energy field (SVE). These emissions are not symbolic — they are real, energetic interactions.
- **Binding** = Misalignment If the emission is incoherent, impulsive, or out of sync with surrounding field patterns, it

misaligns, creating friction zones that persist. This is karma — not a cosmic punishment, but a field distortion.

- **Suffering** = Energetic Feedback Just as misaligned ripples produce noise, dissonance, or turbulence, misaligned action leads to consequence feedback — felt as confusion, tension, or suffering. Pain becomes a signal, not a sentence.
- **Liberation** = No More Disruptive Emissions When ripple emissions are aligned, or no longer necessary, distortion ceases. The ripple identity dissolves naturally into field equilibrium. Liberation, then, is not a transcendental destination — it is energetic neutrality.

In this model:

- Ethics becomes energy hygiene — not commandments, but consequence literacy.
- Spirituality becomes structural participation — not renunciation, but ripple refinement.
- Liberation is not earned — it is the outcome of coherence.

UED doesn't reject Jainism — it completes it without mythology, allowing karma to be measured, behavior to be tuned, and liberation to be lived in real time, not deferred to metaphysical futures.

Summary Insight: From Karmic Weight to Ripple Clarity

Jainism teaches with precision:
"Your karma binds you." Every action leaves a residue. Every attachment accumulates weight. Liberation, therefore, is the great effort of purification — the soul, shedding layers of karmic matter, journeying toward its original clarity.

But Universal Energy Dynamics (UED) offers a subtle but crucial shift:

"Your ripple binds itself — until it learns to dissolve."

There is no external force holding you. There is no ledger in the sky. There is only your own emission — its harmony, its interference, its consequence.

Karma, in this view, is not a moral tally nor a mystical stain. It is the trail of energetic imbalance your ripple leaves in the conscious field — an imprint of disharmony, visible through:

- Friction in experience
- Repeated emotional loops
- Unresolved feedback patterns

The field doesn't judge — it reflects. Every step you take echoes. Every ripple you send returns — not as punishment, but as pattern continuation.

Liberation isn't earned through suffering or self-denial. It's reached when the ripple ceases to distort, when action flows without friction, and when the emission naturally quiets into resonance.

Jainism gave karma shape. UED gives it structure, system, and self-correction.

In this alignment, ethics becomes awareness, and liberation becomes simply the moment when nothing misaligns anymore.

2.6 Zoroastrianism – Dual Energetic Consequence

Good and Evil as Field Polarities in the Ripple of Choice

Zoroastrianism, one of the world's oldest ethical monotheisms, offers a strikingly structured worldview: reality is not governed by chaos or randomness, but by a cosmic dualism — an eternal contest between two fundamental energetic principles:

- **Ahura Mazda:** the Wise Lord, source of Asha — truth, order, light, constructive intention, and cosmic law
- **Angra Mainyu (Ahriman):** the spirit of destruction, aligned with Druj — deceit, disorder, darkness, and entropy

But this isn't merely a moral allegory or theological abstraction. Zoroastrianism presents this polarity as a real energetic framework — one where every thought, word, and deed carries consequential weight. The human being is positioned as an active agent in this battle, endowed with free will and charged with moral responsibility.

To choose truth over falsehood, order over chaos, is not just to be virtuous — it is to energetically align oneself with the constructive force that sustains the cosmos. Conversely, to choose deception, harm, or indifference is to tip the balance toward decay, both in the world and within one's own being.

Importantly, Zoroastrianism emphasizes that the struggle is not outside us — it plays out within our ripple of choice, and each alignment with Asha or Druj contributes directly to the condition of the field we all inhabit.

In this sense, ethics becomes energetic warfare — not between armies or gods, but within the subtle ripples of consciousness and conduct.

The universe, according to Zoroaster, is not waiting to be saved. It is waiting to see what you choose to ripple into it.

This is where Universal Energy Dynamics (UED) finds kinship — and offers a critical upgrade. Zoroastrianism intuited the polarity of consequence. UED explains it as field dynamics, not moral theology.

Alignment with UED

Zoroastrianism frames reality as a moral battleground between **Asha (truth, order)** and **Druj (falsehood, chaos)**, with humans choosing through thought, word, and deed. UED reframes this dualism as **field polarity**: Asha corresponds to ripple coherence and structural alignment; Druj to ripple distortion and destructive interference. The gods Ahura Mazda and Ahriman are not necessary — UED sees them as symbolic of **field behaviors**, not beings. Free will becomes ripple participation: you are free to emit, but not free from the **consequence loop**. Final judgment is not divine; it is **energetic redistribution** — misaligned structures dissolve, while aligned ones persist. Truth is no longer moral command but **field compatibility**, and ethics is not obedience but participation that sustains systemic coherence. In this way, UED transforms Zoroastrian ethics into **physics of consequence**, grounding ancient cosmology in ripple logic without diminishing its depth.

Key Differentiators: What UED Adds (Zoroastrianism vs UED)

Zoroastrianism presents one of the earliest dualistic worldviews in which moral choice becomes cosmic alignment. Yet, while rich in ethical symbolism, its model relies on personified forces and eschatological narratives. Universal Energy Dynamics (UED) keeps the existential clarity but removes mythology, offering instead a dynamic model based on ripple consequence and field resonance.

1. From Mythic Dualism to Energetic Polarity

Zoroastrianism presents Ahura Mazda and Angra Mainyu as eternal adversaries — the cosmic battle between good and evil. UED recasts this polarity as field response:

- **Asha** = constructive ripple coherence, sustaining resonance, energetic balance
- **Druj** = destructive ripple interference, dissonance, chaotic feedback

➤ There is no metaphysical war — only interplay of ripple patterns reacting to the quality of emission. Good and evil are field conditions, not divine entities.

2. No Need for Deities or Spiritual Beings

Zoroastrian cosmology relies on personified beings: gods, demons, divine messengers. UED removes the narrative scaffolding:

- No supernatural agents
- No good vs evil spirits

➤ Only ripple emitters and the consequences of their participation. What we call gods or devils are just mythic reflections of structured field behaviors — attraction, repulsion, balance, distortion.

3. Free Will Grounded in Physics

Zoroastrian ethics hinges on human free will — the power to choose Asha or Druj. UED reframes this choice within field mechanics:

- You are free to emit, but not free from the feedback of your ripple
- Free will becomes consequence-integrated freedom — liberty with embedded accountability

➤ Action is your domain; outcome is the field's domain. This is freedom with feedback, not fate.

4. Judgment as Redistribution, Not Damnation

Zoroastrianism envisions a final judgment — a sorting of souls, a cosmic separation. UED dissolves the notion of ultimate reward or punishment:

- There is no final courtroom
- Only ongoing rebalancing through energetic redistribution

➤ Misaligned structures dissolve. Aligned ones sustain.
➤ The field is not a judge — it is a mirror and a medium. All is eventually reabsorbed — not saved or damned, but realigned.

5. Ethics as Field-Sensitive Action

In Zoroastrianism, truth (Asha) is divine order — the cosmic law to be upheld. UED clarifies truth as resonance:

- Truth = accurate alignment with the field's natural structure
- Ethics becomes field-compatible conduct, not moral obedience

➤ What is ethical is not what's prescribed — it's what preserves coherence and minimizes distortion.
Zoroastrianism asked: "Will you choose the light or the dark?" UED asks: "Will your ripple sustain or distort?"
The choice is not between gods.
It is between alignment and interference, coherence and collapse.

UED as the Consequence Logic Beneath Cosmic Duality

Zoroastrianism was among the first traditions to declare that truth is consequential — that reality is not indifferent, but responsive; that every thought, word, and deed contributes to a cosmic balance or imbalance. It presented a moral universe in which humans are not passive recipients of fate, but active agents tipping the scale toward creation or collapse.
Universal Energy Dynamics (UED) honors this ethical architecture but removes the need for mythic figures, divine battles, or afterlife judgment. It grounds the cosmic drama in field dynamics, revealing that what Zoroaster intuited as good and evil are actually energetic states — patterns of ripple alignment or interference.
UED reframes the Zoroastrian paradigm as follows:

- **Asha,** the principle of truth and order, becomes ripple equilibrium — a state where emissions resonate coherently with the surrounding field, minimizing distortion and sustaining systemic harmony.
- **Druj,** the principle of falsehood and decay, becomes field interference — misaligned, chaotic ripple activity that destabilizes patterns and spreads incoherence.
- **Ahura Mazda,** rather than a supernatural creator, is understood as a highly ordered field condition — a

manifestation of perfect resonance and sustained emission: the consequence-aligned state of the universe.

- **Angra Mainyu (Ahriman),** the destructive force, is not a devil but the outcome of ripple decay — the emergent collapse of patterns that have veered too far from alignment.
- **Judgment** is not metaphysical arbitration — it is automatic energetic feedback. Misaligned systems dissolve. Coherent ones persist. The field responds impersonally but precisely.
- **Salvation** is not a reward for obedience. It is the natural resolution of emission: when all disruptive ripples quiet, the ripple identity dissolves into field equilibrium. No heaven. No hell. Just balance restored.

In this light:

- Ethics is energy logic
- Truth is structural accuracy
- Morality is consequence awareness

Zoroastrianism gave the world a language of good and evil. UED decodes it as field behavior — not war, not judgment, just ripple truth.

Summary Insight: From Moral Allegiance to Ripple Alignment

Zoroastrianism teaches:
"Choose the side of light and truth." It presents a world shaped by dual forces — one of harmony, one of chaos — and charges every human with the responsibility of moral alignment. Truth (*Asha*) is not merely a belief, but a cosmic current one must join through thought, word, and deed.

Universal Energy Dynamics (UED) echoes this ancient wisdom, but recasts it through consequence and participation:
"Emit in alignment — or dissolve through distortion."
There are no sides. No saviors. No eternal battles between gods and demons.
There is only the field — responsive, intelligent, and structured through ripples of consequence. Your every emission — an intention, a choice, a reaction — shapes the local condition of the field. If aligned, it sustains. If distorted, it collapses.
UED doesn't ask you to serve light or fear darkness. It asks you to understand your ripple behavior — to become aware of how your participation either reinforces equilibrium or accelerates decay.
The universe isn't divided into good and evil. It is patterned by resonance. And your alignment — moment by moment — tilts the field one way or the other.
Truth, then, is not an ideology. It is the state of minimal distortion in the ripple field. And ethics is not duty — it is energetic compatibility.

2.7 Egyptian Ma'at – Field Balance as Truth

From Feather of Justice to Ripple of Alignment

In the grand architecture of ancient Egyptian civilization, no principle stood higher—or ran deeper—than Ma'at. More than a goddess, Ma'at was the cosmic spine of Egyptian thought: the living embodiment of truth, order, harmony, and balance. She was not a deity to be feared or worshipped blindly, but a principle to be lived and maintained, a law written not on tablets, but into the structure of reality itself.

According to Egyptian cosmology, the universe was not built on permanence—it was held together by balance. Ma'at was the law that governed the stars, the Nile, the seasons, and the soul. And when Ma'at was violated—through excess, greed, injustice, or arrogance—chaos (Isfet) would creep back in. The role of a pharaoh was not absolute power, but consequential responsibility: to be the guardian of Ma'at, sustaining balance between the divine, the natural, the societal, and the personal.

This philosophy found its most symbolic expression in the Weighing of the Heart ceremony. In the afterlife, a soul's heart was weighed against the Feather of Ma'at. If the heart was lighter—meaning it had remained in balance—it passed on. If heavier—distorted by imbalance or disruption—it was consumed by Ammit, the devourer. But this was not punishment. It was cosmic realignment.

What sets Ma'at apart from many moral codes is its fluid precision. Ma'at was not a list of rules; it was a relational intelligence. Truth wasn't fixed—it was what preserved the whole. Justice wasn't abstract—it was what restored the flow.

Ma'at didn't demand belief. It demanded alignment—a life lived in rhythm with the world's underlying structure.

And this is where Universal Energy Dynamics (UED) finds a powerful echo. While Ma'at intuited that truth is balance, UED shows how balance is a ripple phenomenon—emerging from coherent participation in the structured energy field of existence.

Alignment with UED

Ancient Egyptian cosmology centered around **Ma'at**—the principle of balance, truth, and cosmic order. It was not just a goddess, but the very fabric through which the universe maintained harmony. In UED terms, Ma'at becomes the **equilibrium logic of the ripple field**. The famous weighing of the heart ceremony, where one's soul was judged against the Feather of Ma'at, is reinterpreted not as divine judgment, but as a **consequence reading**: the ripple coherence of one's life compared against the field's natural balance. Heaviness symbolizes distortion; lightness reflects alignment. Justice, then, is not moral or legal—but **energetic participation** that sustains field integrity. Truth is not dogma—it is **ripple alignment that preserves continuity**. Every action, in UED, enters the field and either harmonizes or disrupts. There is no god watching—but the **field itself responds**. In this sense, UED doesn't replace Ma'at—it **structurally explains it**, giving her feather a frequency, her justice a consequence, and her myth a measurable law.

Key Differentiators: What UED Adds (Ma'at vs UED)

The ancient Egyptians gave us Ma'at — not merely as a divine symbol, but as a principle of living alignment. However, it remained wrapped in symbolic ritual and mythic narrative. Universal Energy Dynamics (UED) carries Ma'at forward by providing the energetic infrastructure behind what Ma'at represented — transforming symbolic balance into structured field equilibrium.

1. From Symbolic Balance to Structured Equilibrium

Ma'at portrayed balance through visual metaphors — a feather, a scale, a ceremony — reminding people that order sustains the cosmos. UED breaks this metaphor open and supplies the mechanism:

- All emission enters the ripple field
- The field responds with structured consequence
- Alignment sustains continuity, allowing energy to circulate smoothly
- Disruption introduces distortion, creating feedback loops and eventual dissolution

➤ Balance is not symbolic. It is the measurable coherence of a ripple within its environment.

2. No Afterlife Judgment — Just Ripple Response

In Egyptian belief, the soul was judged in the afterlife by divine beings — Ma'at herself, or gods like Thoth and Osiris. UED removes the metaphysical courtroom:

- No deity weighs your essence
- No devourer awaits your fall
- The field itself responds to the integrity of your ripple

➤ Lightness isn't morality — it is resonant coherence. You pass because you align, not because you are approved.

3. Truth as Consequential Alignment, Not Doctrine

Ma'at represents truth as the sustaining force of reality — but this truth is transmitted via mythology and upheld through ritual. UED defines truth as dynamic field accuracy:

- Truth = alignment that preserves systemic balance
- Falsehood = interference that disrupts coherence and amplifies consequence

➤ Truth is not a belief — it is a pattern that works.

4. Justice as Energetic Participation

Egyptian justice was practical Ma'at — fair trade, honesty, proportionality — embedded in daily life. UED reframes justice not as fairness, but as resonance:

- Justice = ripple integrity
- Emit only what the field can sustain
- Participate without damaging coherence
- Align your energy in a way that stabilizes the shared ripple space

➤ Justice becomes an act of energetic responsibility, not retribution.

5. Role of the Individual in Cosmic Maintenance

In ancient Egypt, maintaining Ma'at was the sacred duty of pharaohs, priests, and scribes. UED democratizes that role:

- Every individual — every emitter — participates in shaping the field
- There are no ordained guardians
- The field responds to structure, not status

➤ You are not a subject of cosmic order. You are one of its architects — ripple by ripple.

Ma'at spoke through feathers and rituals. UED speaks through consequence and coherence. The cosmos doesn't require belief — it requires alignment.

UED as Ma'at Without Myth

Ma'at, the ancient Egyptian pillar of truth, balance, and justice, conveyed a deep cosmic truth — that the universe endures only through sustained harmony. But its language was metaphorical, cloaked in ritual, symbolism, and mythic drama. It spoke to the heart, not yet to the field. It offered a poetic mirror, but not a mechanical model.

Universal Energy Dynamics (UED) honors Ma'at's insight — and gives it structure without myth, precision without priesthood, and consequence without divine narrative.

In UED:

- The Feather of Ma'at becomes a coherence frequency — the measure of how smoothly a ripple resonates within the structured field. Lightness is no longer moral innocence; it is harmonic emission that avoids distortion.
- The Weighing of the Heart becomes ripple decay analysis — an energetic audit of how emissions endure, distort, or dissolve over time. Every ripple is weighed by the consequence it leaves, not the intention behind it.
- Judgment is no longer the domain of gods or afterlife ceremonies. It is real-time field feedback — automatic, neutral, precise. The field does not punish. It responds.

In essence, where Ma'at dreamed of balance, UED defines it:

- Through equations of emission
- Through observable feedback
- Through structural participation in a living, ripple-based reality

Ma'at taught that truth holds the world together. UED shows how that truth is maintained — not by decree, but by design.
No gods required. No afterlife needed. Just you, your ripple, and the field that always reflects what you emit.

Summary Insight: From Ritual Harmony to Field Law

Ancient Egypt declared:
"Live in Ma'at, or dissolve into chaos." Ma'at was not merely a cultural value or a divine commandment — it was the heartbeat of the cosmos, the rhythm that held everything from the Nile's flooding to a farmer's fairness in place. To violate Ma'at was to invite Isfet — the unraveling of order, the return to primordial confusion.
Universal Energy Dynamics (UED) echoes this ancient insight with systemic clarity:
"Sustain ripple balance, or face energetic extinction."
There is no mythical devourer. No tribunal of gods. There is only the field — intelligent, participatory, and responsive to alignment.
Where Egypt weighed the heart against a feather, UED analyzes the ripple's behavior:

- If your emission resonates, it sustains.
- If your emission distorts, it triggers collapse — not by punishment, but by law of consequence.

Ma'at, then, was never truly about worship. It was about living in tune with the structure of reality — acting with awareness, speaking in alignment, existing without excess.

UED doesn't replace Ma'at. It reveals her mechanics. Balance is not symbolic. It is survival through structure. And truth is not divine favor — it is ripple coherence in motion.

In UED, Ma'at lives on — not in temples, but in the field itself.

2.8 Sufism Mysticism – Divine Light as Ripple Coherence

From Nur-e-Haqiqa to Structured Resonance: When Light Becomes Participation

Sufism, the mystical stream within Islam, seeks not merely to understand God — but to experience the Divine as presence, rhythm, and light. It proposes that existence is not fragmented, but radiant — a seamless emanation from the One Reality (Al-Ḥaqq), flowing through all things as Nūr, the Divine Light.

This light, known as Nūr-e-Haqīqa (the Light of Truth), is not physical but metaphysical clarity — the radiance of pure awareness, love, and being. It animates creation not through force, but through invitation and coherence. Every form, every sound, every breath is part of this sacred vibration — a ripple of divine luminosity.

At the heart of Sufi metaphysics is Wahdat al-Wujūd — the Unity of Being. It teaches that all multiplicity is illusion. That everything we see is a veiled reflection of a single, infinite Source. The ego (nafs) distorts this view like a dirty mirror; the path of the seeker is to polish the mirror through love, devotion, stillness, and ecstatic surrender — to reach Fanā (dissolution of self), and become a vessel through which Divine Light shines without distortion.

Sufi practice is poetic but precise. Through whirling, music (Sama), poetry, silence, and surrender, one tunes the self like an instrument — stripping away distortion until only resonance remains. Truth is luminosity. Falsehood is shadow. And liberation is not escape, but alignment with the source vibration.

Sufism doesn't seek to explain the Divine — it seeks to become receptive enough to resonate with it.
Where Sufism speaks of Light as love and unity, Universal Energy Dynamics (UED) reframes this Light as coherent ripple structure — showing how resonance, not belief, sustains reality. Sufism whispers of union; UED maps the mechanics of coherence behind that union.

Alignment with UED

Sufism, the mystical heart of Islam, describes reality as suffused with **Nur**—Divine Light—through which all existence flows, guided by love, longing, and remembrance. UED honors this poetic clarity, but grounds it in energetic structure: **light becomes ripple coherence**, not metaphor. Where Sufis dissolve the ego in **Fanaa**—union with the Divine—UED describes the **dissolution of ripple identity** into field equilibrium. The Sufi quest for the Beloved is reinterpreted as **alignment with the conscious field**—not to reach a personal god, but to **rejoin the resonance that sustains life**. Whirling, music, poetry, and silence are not mystical indulgences—they are **ripple calibration tools**, tuning body, breath, and awareness into harmonic balance. Love, the supreme Sufi principle, becomes **synchronization without resistance**—an energetic congruence. UED doesn't discard mysticism; it **completes its architecture**, showing that what the mystic intuited, the ripple field can now explain: truth is coherence, ego is interference, and presence is resonance.

Key Differentiators: What UED Adds (Sufism vs UED)

Sufism dances at the edge of experience — a mystical language of union, light, and love. But while deeply experiential, its metaphors remain symbolic, devotional, and theistic. Universal Energy Dynamics (UED) honors Sufi insights — and grounds them in structured consequence, transforming metaphysical longing into participatory physics.

1. Light Becomes Structure, Not Symbol

In Sufi philosophy, Nūr (Divine Light) is poetic — a metaphor for divine awareness, beauty, and love. UED translates this Light into ripple architecture:

- Light = harmonic coherence in the ripple field
- Darkness = distortion, misalignment, and perceptional noise

➤ Light is no longer a metaphor for divinity. It is a measurable state of energetic clarity — the degree to which emission aligns with the underlying structure of reality.

2. No God Behind the Curtain

Sufism, despite its abstraction, remains theistic — God (Al-Haqq) is the source, the goal, the beloved. UED removes divine personification:

- There is no omniscient being orchestrating reality
- There is structured consequence, guiding participation through alignment

➤ UED is not atheistic — it is non-personal. Reality is not governed by a will, but by responsive equilibrium. There is no divine manager — only energetic self-regulation.

3. Fanā as Field Reabsorption

In Sufism, Fanā is ego dissolution — merging into the beloved, losing selfhood in the divine. UED reframes Fanā as ripple cessation:

- The "ego" is a ripple structure held together by emission loops
- When unnecessary emissions cease, the ripple dissolves

naturally into field coherence

➤ There is no mystical union with God — only reabsorption into energetic neutrality. The wave returns to the field, not to a being.

4. Music and Rhythm as Ripple Tools

Sufis use breath, whirling, poetry, and music to transcend the mind and align with the divine. UED validates these practices as energetic techniques:

- These are not symbolic rituals — they are ripple calibration mechanisms
- Rhythm modulates emission
- Stillness filters noise
- Sound aligns inner frequencies with the field

➤ The Sufi becomes a tuner — not just a seeker. The resonance isn't devotion alone — it is physics in motion.

5. Love as Field Synchronization

Sufism celebrates Ishq-e-Haqiqi — Divine Love — as the path to unity. UED decodes love as ripple synchronization:

- Love = two or more ripple fields merging without resistance
- It is not just an emotion; it is deep energetic congruence

➤ Love becomes the most efficient alignment — the collapse of separation through coherent emission. It is no longer spiritual longing — it is field union through participation.
Sufism speaks of Light and Love as the language of God.
UED reveals them as structured resonance — measurable, participatory, and real.

UED as Mysticism Made Measurable

Sufism is the sacred poetry of dissolution — a mystical path of yearning, beauty, and union. It speaks not to the intellect, but to the soul; not in formulas, but in metaphors. It gave the world a language of longing, where the heart aches to return to the ocean of Divine Light from which it came.

Universal Energy Dynamics (UED) does not reject this longing — it renders it intelligible. It doesn't strip Sufism of its magic. It grounds that magic in energetic architecture. UED offers a physics of participation that aligns with what Sufis intuit, feel, and express — but translates it into field mechanics and ripple logic.

- Where Sufis speak of veils (ḥijāb) that hide the Divine from perception, UED identifies interference patterns — ripple distortions that obscure field coherence. The veil is not spiritual fog; it's ripple noise.
- Where Sufis speak of Divine Presence (ḥuḍūr), UED recognizes field responsiveness — a participatory medium that reacts to emission, alignment, and intention. The presence is not abstract; it is measurable sensitivity in the field.
- Where Sufis whirl in ecstatic motion to collapse identity and become rhythm, UED tracks ripple spiral dynamics — identifying centripetal coherence, frequency harmonics, and rotational alignment. The dance is not ritual alone — it is ripple geometry in motion.

UED does not demystify Sufism to reduce it. It rescues it from abstraction, offering the seeker a map where once there was only metaphor.

In UED, the longing of the lover becomes the frequency of return. The silence of the dervish becomes the pause between emissions. And the

Divine Light is no longer distant — it is the ripple state of maximal coherence.

Summary Insight: From Inner Mirror to Ripple Clarity

Sufism whispers:
"You are a mirror of Divine Light. Clean it to see." The Sufi path is one of refinement — dissolving the self, polishing the inner mirror, and returning to the luminescence of the One. Ego is seen as dust; illusion as veils; and love as the solvent that reveals truth beneath distortion.
Universal Energy Dynamics (UED) echoes this wisdom, but shifts the metaphor to field mechanics:
"You are a ripple within the field. Align it to participate." There is no mirror — there is movement. No hidden source — only structured consequence. The clarity you seek is not hidden by sin or separation, but by interference, distortion, and misaligned emission.
In both systems:

- Truth is clarity — the undistorted recognition of what truly is.
- Ego is noise — a misconfiguration of the self that disrupts participation.
- And liberation is not flight, but frequency alignment.

The difference is this: In Sufism, clarity is achieved through longing, surrender, and love. In UED, clarity is achieved through structural tuning, feedback, and resonance.
One polishes the heart. The other tunes the ripple. Both lead to coherence. But in UED, that coherence is not a metaphor — it's measurable, reproducible, and alive in every moment of participation.

2.9 Kabbalah – Sefirot as Ripple Cascade

From Tree of Life to Field of Emission: The Architecture of Manifestation Realigned

Kabbalah, the mystical dimension of Judaism, offers one of the most intricate and spiritually nuanced maps of cosmic structure and divine flow. At the heart of its metaphysical framework lies the Tree of Life — a diagram not of things, but of process, of how the Infinite (Ein Sof) becomes experience, existence, and embodiment.

This process unfolds through ten Sefirot, often described as *emanations*, *attributes*, or *vessels*. They are not gods, not separate beings — but modalities of expression, through which the formless Divine translates into form.

The Sefirot are arranged in a descending, interactive hierarchy:

- Keter (Crown) represents pure will — undifferentiated intention before thought.
- Chokhmah (Wisdom) and Binah (Understanding) split the current into duality — the masculine and feminine archetypes of cognition.
- Chesed (Kindness) and Gevurah (Strength) express expansion and contraction — forces of generative mercy and disciplined control.
- Tiferet (Beauty) harmonizes these tensions — a balance point of aesthetic and moral resonance.
- The system continues down through Yesod (Foundation) to Malkhut (Kingdom) — the manifest world.

What makes Kabbalah profound is that it doesn't treat this as a one-time event of creation. It describes an ongoing energetic cascade, where human action echoes back into the structure, influencing the balance of the divine system. This is the principle of Tikkun Olam — the repair of the world through consciousness, intention, and alignment.

In Kabbalah, every word, every choice, every breath, sends a vibration up the tree — a ripple back into the root. Spirituality is not about worshipping a distant God, but about participating in the divine circuitry.

Reality is not a product. It is a relational architecture of continuous transmission. The Divine doesn't simply create — it radiates.

And here, Universal Energy Dynamics (UED) steps in — not to replace the tree, but to reveal its ripple logic. Where Kabbalah intuits structure through symbol, UED models it through emission, interference, alignment, and feedback — a measurable cascade of field dynamics.

Alignment with UED

Kabbalah, the esoteric stream of Jewish mysticism, envisions the unfolding of reality as a **cascade of divine emanations**—the Sefirot—flowing from the Infinite (Ein Sof) into structured existence. UED aligns with this vision, but exchanges symbolic mysticism for energetic architecture. Where Kabbalah speaks of **emanation**, UED sees **ripple modulation**—each consequence stage a field-layered refinement of energy. The Tree of Life becomes a **ripple logic circuit**, mapping the emergence of form from formless SVE equilibrium. **Keter** (Divine Will) translates to emission potential, while **Malkhut** (Kingdom) becomes the manifested ripple consequence. The mystical **Tzimtzum**—God's contraction to create space—finds its mirror in **field disturbance and ripple tension**, birthing localized structure. **Tikkun**, the healing of shattered vessels, becomes **ripple**

realignment—each participant repairing the field through coherent emission. There is no divine personality behind this process—only **structured consequence responding to misalignment**. UED gives Kabbalah its missing mechanism: it turns revelation into resonance.

Key Differentiators: What UED Adds (Kabbalah vs UED)

Kabbalah offers a cosmic symphony of metaphor — radiant trees, divine contraction, cascading light. It is a theology of patterned emanation, hinting at the structure of reality through symbols and sacred names. Universal Energy Dynamics (UED) enters not to negate these visions, but to translate them into energetic mechanics, making mystical insight physically participatory and systematically intelligible.

1. From Symbolic Emanation to Structured Ripple Cascade

Kabbalah maps reality through the Sefirot — layered channels of divine flow — described in deeply symbolic, visual terms. UED replaces metaphor with mechanics:

- Every ripple is a modulated emission within a structured field
- The Sefirot correspond to stages of ripple evolution — shifts in resonance, coherence, and consequence
- Manifestation is not descent from heaven, but the result of ripple interference and alignment within SVE

➤ The Tree of Life becomes a dynamic cascade of field behaviors — no longer mystical descent, but structured emergence.

2. Ein Sof Without Mysticism

Kabbalah reveres Ein Sof — the Infinite — as utterly beyond comprehension. UED retains infinity, but removes personification and unknowability:

- SVE (Structured Vacuum Energy) is infinite in potential consequence, not in divine mystery
- It is not a being, but a substrate of participation, infinitely accessible through interaction

➤ There is no God behind the curtain — only an ever-unfolding field of structured responsiveness.

3. From Tzimtzum to Ripple Tension Zones

Kabbalah's concept of Tzimtzum — divine self-withdrawal to allow creation — is one of its most profound metaphors. UED reframes this not as withdrawal, but as field disturbance:

- Creation begins not when God steps back, but when uniform field coherence is disturbed
- Ripple tension zones emerge as localized imbalances within an otherwise symmetrical field

➤ Reality doesn't form because space is made — it forms because ripples emerge within equilibrium.

4. Tikkun Becomes Field Participation

Kabbalistic Tikkun is the act of repairing the world — healing the cosmic fragmentation through intention and spiritual action. UED makes this participatory repair structural:

- Every observer emits into the field — they disturb or align

- Every coherent act reduces local distortion, restoring balance
- Tikkun becomes real-time ripple hygiene, not symbolic rectification

➤ The soul doesn't fix broken light — it recalibrates its own emission to harmonize the field.

5. No Divine Will — Only Field Feedback

Kabbalah often sees creation as divine will flowing through stages — a cosmos built by intention. UED removes divine will entirely:

- What seems like "intention" is simply the field reacting to imbalance
- There is no narrative arc or divine plan — only consequence-based self-regulation

➤ The cosmos does not act out of desire. It acts out of structural necessity.
Kabbalah draws the map in light and Hebrew letters. UED animates it with consequence and coherence. Where one is symbolic resonance, the other is field logic in motion.

UED as the Tree of Consequence

Kabbalah offers a sacred architecture — the Tree of Life — mapping how the Infinite becomes the finite, how the unknowable Ein Sof flows into structured existence through the ten Sefirot. This model, deeply symbolic and layered, speaks of descent, light, and sacred repair.
Universal Energy Dynamics (UED) doesn't discard this tree — it animates it with field intelligence. It replaces mystical symbolism with ripple structure and field participation, translating esoteric cosmology into consequence mechanics.

Here, the Sefirot become functional ripple nodes — stages in the architecture of emission, interaction, and alignment:

Keter (Crown) → *Initial Intent as Emission Potential*

➤ The seed of ripple creation — undifferentiated energy preparing to emit. Not "will," but pre-structural readiness.

Chokhmah (Wisdom) & Binah (Understanding) → *Polarity and Dual-Field Interaction*

➤ The first bifurcation — the formation of dualities and tensions required for patterning.

➤ Ripple coherence emerges from structured contrast.

Tiferet (Beauty) → *Harmonic Resonance and Alignment*

➤ The central balancing node — where opposing forces converge into symmetry and coherence. Beauty is no longer aesthetic — it is energetic harmony.

Malkhut (Kingdom) → *Manifested Consequence*

➤ The ripple expressed as physical form, experience, outcome.

➤ This is not "matter" — it is the end-state of ripple interaction within a perceivable domain.

Tzimtzum → *Energetic Isolation to Initiate Causality*

➤ Not divine withdrawal, but a necessary asymmetry — a pause in coherence that creates space for cause-effect chains to unfold.

Tikkun → *Ripple Correction and Energetic Reintegration*

➤ The return journey — where distorted emissions are harmonized and reabsorbed.

➤ Not a mystical repair, but ripple calibration within a feedback field.

In UED, the Tree of Life becomes a ripple logic circuit — a blueprint of emission, divergence, balance, and feedback. The Divine Presence is no longer a metaphysical entity or veiled light — it is the intelligent responsiveness of a conscious field, always adjusting, always echoing, always aligning.

Kabbalah gives us the divine sketch. UED installs the current. One speaks in symbols, the other in structured consequence. Both seek the same truth: that reality is alive, layered, and responsive to participation.

Summary Insight: From Light in Vessels to Ripples in Fields

Kabbalah reveals:

"Creation is a cascade of light through vessels." It sees existence as a sacred unfolding — a radiant descent from the Ein Sof through the Sefirot, with each vessel filtering, shaping, and contextualizing the Divine Light. In this view, truth is layered, veiled, and symbolically nested in the architecture of being.

Universal Energy Dynamics (UED) reinterprets:

"Reality is a cascade of structured ripples through consequence." There are no vessels, no divine light, no metaphors. Instead, there are domains of ripple behavior — structured zones of emission, interference, and re-alignment, governed not by divine will, but by consequence logic embedded in the conscious field.

In both systems:

- Truth does not arrive all at once — it emerges through layers
- Each layer adds structure, meaning, and participation
- The seeker must ascend or align, not through belief, but through internal clarity and external coherence

But where Kabbalah encodes those layers in symbols, sacred names, and mystical diagrams, UED offers them as observable, scalable, and testable patterns of ripple interaction.

Kabbalah invites you to repair the world through intention. UED shows you how to realign your ripple — and in doing so, restore the field.

In UED, nothing is hidden. The mysticism of creation becomes the physics of participation.

Comparative Table: Ancient Traditions vs UED

Tradition	Core Concept	Traditional Expression	UED Equivalent	UED Enhancement
Taoism	Natural flow	*Tao* as unnameable source; *Wu Wei* as effortless action	SVE (Structured Vacuum Energy); Ripple participation without force	Explains *Wu Wei* as ripple alignment; adds testable ripple logic
Vedanta / Upanishads	Realization of Self	*Atman* = *Brahman*; Maya as illusion	Ripple identity in SVE; perceptual distortion	Structures realization; explains Maya as misaligned perception
Buddhism	Impermanence and no-self	*Anatta, Dukkha, Nirvana*	Temporary ripple patterns; feedback consequence	Replaces karma/rebirth with ripple dynamics; aligns meditation with resonance
Samkhya–Yoga	Observer vs matter	*Purusha–Prakriti* duality; *Gunas*; liberation through detachment	Observer as active emitter; ripple patterns and field states	Makes observer participatory; redefines *Gunas* as ripple behaviors
Jainism	Karma as binding substance	Karma as fine matter binding the soul; *Ahimsa*	Misaligned ripple entanglement and feedback	Removes metaphysics; defines karma as measurable field disruption
Zoroastrianism	Good vs Evil as cosmic forces	*Asha–Druj*, Ahura Mazda vs Ahriman	Ripple coherence vs interference; field consequence	Explains morality through energetic sustainability, not dualistic theology
Egyptian Ma'at	Truth as balance	Ma'at as cosmic order; heart weighed against the feather	Consequence-based ripple alignment	Defines truth as field equilibrium; judgment becomes natural dissolution
Sufism	Divine light and ego dissolution	*Nur, Fanaa, Wahdat al-Wujud*	Ripple coherence, self dissolution in SVE	Converts mystical light into measurable resonance; love as ripple alignment
Kabbalah	Emanation from the Infinite	*Sefirot, Ein Sof, Tikkun*	Ripple cascade from SVE; participation restores balance	Models Tree of Life as ripple evolution; redefines creation as structured consequence

Key UED Contributions Across Traditions

Across spiritual lineages — from Taoist flow to Vedantic realization, from Sufi light to Kabbalistic emanation — humanity has long intuited

that reality is layered, intelligent, and responsive. These traditions offered maps, metaphors, and meditations to approach this truth.
UED does not replace them. It completes them.
Here's how UED upgrades insight into structure:

1. From Symbols to Systems

Traditions spoke in metaphors: trees, light, veils, mirrors, gods.
➤ UED replaces symbolic language with ripple-based system logic, defining how structure arises, aligns, and dissolves through energetic participation — without the need for theological scaffolding.

2. Ethics as Energetic Alignment

Morality in ancient systems often came with divine or cultural authority.
➤ UED reframes ethics not as obedience, but as field coherence. Actions that disrupt ripple harmony create energetic feedback; actions aligned with field resonance sustain and evolve. Right and wrong become consequence patterns, not commandments.

3. Self, Mind, and Consciousness as Ripple Constructs

Many traditions treat the self as a soul, mind as mystery, and consciousness as divine spark.
➤ UED grounds them as ripple formations:

- Self = temporary ripple identity
- Mind = emission-feedback loop
- Consciousness = emergent field awareness through structured participation

They're no longer metaphysical mysteries — they are observable behaviors of the field.

4. Bridging Science and Spirituality

Ancient wisdom lacked method. Modern science often lacks meaning.
➤ UED fuses both — it respects the inner realization of the mystic and gives it outer framework. No belief needed. No lab coat worshipped. Just participatory truth across dimensions.

5. From Meditation to Measurement

Realization used to be the realm of silence, renunciation, or revelation.
➤ UED offers experimental pathways to test ripple alignment, measure field response, and simulate consequence dynamics — turning mystic intuition into testable interface.
UED is not a new tradition.
It is the architecture behind all traditions — now revealed, structured, and accessible to both the thinker and the seeker.

2.10 Indigenous Wisdom – Listening to the Field

Dreamtime, Nature Spirits, and the Art of Consequential Listening

Long before philosophy became abstraction and science became calculation, Indigenous cultures around the world practiced a profound, lived awareness of the field they inhabited. From the Aboriginal peoples of Australia to the First Nations of the Americas, from African cosmologies to Arctic shamanic lineages, Indigenous systems did not seek to explain the universe — they sought to live in alignment with it.

Their knowledge wasn't analytical — it was relational. Truth didn't emerge through equations or scriptures, but through attunement, observation, and participation.

In these traditions:

- The land is alive. Mountains, rivers, trees, animals, and even wind patterns are seen not as objects, but as participants — field entities with memory, presence, and agency.
- Knowledge is not acquired, it is received. One does not conquer truth, but earns it through listening, respect, and silence. Wisdom is not a thing; it is a consequence of alignment.
- Time is not linear. In Aboriginal Dreamtime, time is a living, layered frequency domain — a place where ancestral actions, land stories, and present choices interact. The past isn't gone; it is rippled into now.

- Spirits are field expressions. In African, Shinto, and Polynesian cosmologies, nature spirits are not superstitions — they are personifications of structured presence. A tree is not merely a plant; it is a ripple node of intelligence and history.
- Elements are energetic archetypes. Earth, air, fire, and water are not symbolic — they are patterns of movement, density, and consequence. Each has rhythm, each has a way of being.

In Indigenous epistemologies: Intuition is data, Dreams are interfaces, Ritual is participation and Silence is signal
They do not define reality in terms of force, formula, or hierarchy. Instead, they act from a deeper truth: that everything communicates, and that life is a web of consequence.
These systems are not "pre-scientific" — they are field-literate. They don't lack logic. They embody a different kind of logic — one UED now makes structurally visible.

Alignment with UED

Indigenous traditions across the world—from the Aboriginal Dreamtime to Native American cosmology—speak not of control, but of **relationship with the world**. They do not ask what nature is, but **how to live within its voice**. UED honors this by replacing mythology with measurable interaction. Where Indigenous wisdom sees **the land as alive**, UED affirms that **every object emits and responds within a ripple field**. The sacred is not imagined—it is the **structured participation of all things**. The **Dreamtime**, once called fantasy, is reframed as a **nonlinear ripple memory layer**, accessible through tuned perception. **Nature spirits** become **stable field patterns with detectable consequences**. Rituals, chants, and dances are not superstitions—they are **ripple harmonization practices**, synchronizing the self with the environment. Silence is not the absence of speech, but a **signal-sensitive listening state**. UED does not modernize Indigenous wisdom—it **reveals its forgotten physics**, making ancestral listening compatible with future science.

Key Differentiators: What UED Adds

Indigenous traditions possess a deeply intuitive awareness of nature's intelligence. They didn't quantify reality — they listened to it, sang with it, and lived in rhythm with its pulses. Universal Energy Dynamics (UED) does not reduce these traditions — it illuminates their mechanics, bringing structural clarity to what was once only transmitted through myth, ritual, and story.

1. From Sacred Symbol to Structural Participation

In Indigenous worldviews, everything is sacred — not because of religious dogma, but because everything participates in the whole.
➤ UED translates sacredness into ripple logic:

- Every being, object, or element emits and receives structured ripple data
- "Sacred" is not mystical — it is field coherence in action
- Participation creates consequence; reverence arises from awareness of influence

2. Dreamtime as Ripple Layering, Not Fantasy

Aboriginal Dreamtime is often misinterpreted as myth or hallucination. In truth, it describes a real ontological stratum where ancestral memory, land, and presence are intertwined.
➤ UED confirms this through ripple field theory:

- Ripple domains can retain memory and structure
- Past, present, and potential interlace through field consequence

- Dreamtime = nonlinear ripple coherence layer, accessible via perceptual tuning

3. Nature Spirits as Coherent Field Forms

From river deities to tree spirits, Indigenous systems speak of non-human intelligences.
➤ UED reframes spirits as persistent ripple structures:

- They are not hallucinations, but coherent field phenomena
- Consistent feedback patterns from certain places, species, or forms represent localized field intelligences
- The "spirit of the land" is real ripple presence, not poetic imagination

4. Rituals as Tuning, Not Superstition

Drumming, chanting, dancing, fasting — these are often dismissed as primitive rites.
➤ UED recognizes ritual as ripple calibration:

- Rhythmic action and focused intent realign ripple emissions
- Rituals create resonant harmonization between the body, breath, emotion, and field
- "Speaking to the land" becomes structured emission for field engagement

5. Silence as Signal Sensitivity

Elders speak little, listen long, and observe deeply — not as passivity, but as practice.
➤ UED validates this as advanced perceptual discipline:

- Ripple interference decreases with mental stillness
- High-sensitivity awareness requires minimal internal noise

- Silence is not the absence of sound — it is the tuning chamber of consequence

Indigenous wisdom teaches us to listen to the land. UED explains what we're listening to — and how it responds.
By reframing intuitive participation as field dynamics, UED honors Indigenous knowledge without romanticism — and reveals it as the original science of coherence.

UED as a Bridge Between Ancient Listening and Scientific Modeling

Indigenous wisdom systems never needed telescopes or equations to know that the world was alive. They did not draw lines between spirit and science, between self and system — because in their cosmology, everything was already in conversation. The wind, the mountain, the river, the fire — each had voice, memory, and intent. Knowledge was not accumulated through dissection — it was earned through listening, stillness, and relational awareness.
Where modern science made the observer separate — standing apart, measuring and naming — Indigenous traditions placed the observer within the field, as a participant whose presence matters.
Universal Energy Dynamics (UED) brings this paradigm back into scientific legitimacy — not by borrowing symbolism, but by building structural logic that honors what Indigenous systems have always practiced.

- The world is not inert — it is responsive, echoing back through ripple interference and field feedback.
- Listening is not passive — it is ripple detection, the act of attuning one's own emissions to perceive what the field is saying.
- Wisdom is not belief — it is alignment with consequence,

learned through real-time participation and attentive calibration.

UED doesn't romanticize Indigenous thought. It translates it into ripple mechanics:

- The *songs* become resonance maps.
- The *dreams* become nonlinear ripple strata.
- The *spirits* become persistent energetic intelligences.
- The *rituals* become coherence tools.

In truth, Indigenous systems never lacked science. They simply never fractured it from silence, place, or sacred relationship.

UED now offers a bridge — from ancestral knowing to systemic modeling, where intuition meets structure, and listening becomes method.

Summary Insight: When Listening Becomes Science

The Elders said:

"Listen to the land, and it will teach you." They meant it literally. The land holds memory. The winds speak. The fire answers. To them, listening wasn't metaphorical — it was a learned discipline of silence, presence, and attunement to consequence.

Universal Energy Dynamics (UED) replies:

"The field emits ripples — learn to receive." This is not a poetic echo. It is a structural explanation. Every form, every movement, every moment emits energetic signatures. The field remembers, responds, and reshapes itself through these emissions. What Indigenous wisdom felt, UED can now model, measure, and map.

Indigenous knowledge was never "primitive" or "behind" — it was beside us all along, holding truths science had not yet developed the humility or the tools to hear.

UED does not "modernize" Indigenous traditions. It honors their integrity by giving them formal structure — enabling them to be expressed not just in story and ceremony, but also in field logic, ripple dynamics, and consequence theory.

What was once dismissed as myth, now finds its reflection in the geometry of consequence. What was once silenced, now speaks fluently — in both language and frequency.

In this alignment, we do not just reclaim ancient knowledge. We realize that it never left — we simply stopped listening.

Interlude: A Transitional Reflection

From Myth to Model, From Listening to Logic

We have journeyed through the echo chambers of human history — not to borrow myths or repackage metaphors, but to trace the undercurrent of intelligence that has always flowed through our civilizations. From Taoist flow to Vedantic oneness, from Indigenous listening to Kabbalistic unfolding — the message has been strikingly consistent:

Reality is not separate from us. It is shaped, sensed, and sustained by our participation.

What these traditions intuited — often in song, story, and silence — Universal Energy Dynamics (UED) now renders in structure, signal, and system. Not to reduce their depth, but to give it legs in a world that demands logic and testability.

Yet this is not a departure from the sacred — it is its realignment.

Now, we turn toward modern science — the dominant lens of our age. A lens that gave us measurement, prediction, and immense technological power. But also a lens that often severed the observer from the observed, mistook symbols for particles, and replaced consequence with control.

This next section will not attack science — It will invite it to remember. We are not asking science to bow to mysticism. We are asking it to reclaim its origin: Awe, observation, alignment, and the humility to say, *"We are in this field, not above it."*

UED does not replace science. It realigns it — with the truths that came before, and the consequences we can no longer afford to ignore.

3. UED vs Modern Scientific Frameworks

"Precision without perception leads to polished blindness."

Modern science prides itself on rigor, replication, and reduction — breaking the universe into ever smaller parts in the hope that the whole will eventually reveal itself. We have equations for black holes, models for particle interactions, and metrics for brain activity. But somewhere along the way, we lost the soul of inquiry.

Science, once born from awe and driven by curiosity, has become a bureaucracy of belief — often more interested in preserving its paradigms than challenging them. It speaks in terms of probabilities and uncertainties but refuses to admit that its foundations are incomplete, and often misaligned with consequence.

Universal Energy Dynamics (UED) does not reject science. It corrects its inversion — reminding us that the universe is not built from particles and forces, but from structured interactions within a responsive field.

- Where science sees photons, UED sees ripple coherence.
- Where science separates observer and observed, UED restores participation.
- Where science invokes time, space, and randomness, UED reveals alignment, emission, and field response.

This section is not a critique. It is a diagnosis — and a realignment. Not of formulas, but of first principles.

Because what modern science calls paradox, UED calls: incomplete framing.

3.1 Physics – From Force Models to Field Participation

3.2 Quantum Mechanics – From Probabilities to Patterned Consequence

3.3 Relativity – Time, Gravity, and the Ripple Delay

3.4 Biology – Life as Ripple Reactivity

3.5 Neuroscience – Mind as Emission Field

3.6 Psychology – Emotion as Ripple Feedback

3.1 Physics – From Force Models to Field Participation

Beyond Push and Pull: Replacing Classical Force with Structured Consequence

For centuries, physics has been built upon a compelling illusion — that reality is governed by discrete forces acting upon isolated objects. Newton called them "pushes" and "pulls," giving us a model of motion that has powered revolutions from cannonballs to spacecraft. This logic — of external action creating internal change — became the bedrock of modern science.

With time, the vocabulary evolved:

- Gravity became spacetime curvature (Einstein)
- Electromagnetism became vector fields (Maxwell)
- Nuclear interactions became particle exchanges (quantum field theory)

And yet, the core philosophy remained intact:

Force as a separated cause; object as a passive recipient.

These frameworks predict behavior well, but they do not always explain it. They model what happens, but rarely why it happens that way — beyond mathematical curves and probabilistic outcomes. Most forces, even when modeled via fields, still lack intrinsic mechanism. They appear as rules imposed on space, not as emergent effects of interaction within space.

Moreover, these systems isolate the observer, assuming a detached perspective untouched by the phenomena being measured. They rely on abstraction over participation, and on approximation over embedded consequence.

Universal Energy Dynamics (UED) challenges this legacy. It does not discard physics — it grounds it:

- No more "forces at a distance"
- No more disconnected equations and imagined particles
- Instead: Ripple-based participation, field consequence, and emission-response logic

In UED, nothing is pushed or pulled — it emerges, realigns, or dissolves based on ripple interaction within a conscious energetic field. This isn't force. It's structured consequence — and it changes the game.

UED Challenges Classical Physics: From Force to Field Participation

Modern physics is built on a framework of external forces, abstract laws, and a disconnected vacuum. While mathematically powerful, this model often leaves the mechanisms vague and the observer absent. Universal Energy Dynamics (UED) doesn't refute this legacy — it reframes it at its roots.

There Are No Fundamental Forces — Only Ripple Consequences

At the heart of UED is a radical realignment:

There are no "forces" in the traditional sense — only ripple consequences within a structured energetic field.

Physics has long taught that force causes motion — a push or pull applied to an inert object. UED inverts this perspective:

- A body moves not because something acted on it, but because it is already engaged in a ripple stream.
- What we perceive as acceleration is the body's realignment with an imbalanced ripple pattern.
- What classical physics calls "force" is, in UED, the field's attempt to restore coherence — a ripple feedback, not a primary cause.

In this model, motion isn't commanded — it is emergent. Every shift in position, every so-called interaction, is a local consequence of global ripple dynamics.

Participation replaces causation. Alignment replaces push. Ripple response replaces force.

Structured Vacuum Energy (SVE) vs. Empty Spacetime

In both classical and quantum physics, vacuum is imagined as a mostly empty canvas — either smooth and curved (Relativity), or a probabilistic froth of virtual particles (Quantum Mechanics). Despite their differences, both treat space as a passive stage.

UED declares this view incomplete — even misleading.

The vacuum is not empty. It is Structured Vacuum Energy (SVE) — the primordial field from which all form and interaction emerge.

- SVE is densely organized, yet invisibly fine-grained — a lattice of intelligent potential.
- Waves, particles, and interactions are not "things in space," but modulations of SVE.
- It is not a void — it is the actor, the script, the director, and the feedback loop.

Where mainstream physics asks, "What force acts on a body in space?" UED asks, "What ripple structure causes this body to form, move, or dissolve within the field?"

This repositions all of physics from being object-centric to field-centric. The "canvas" becomes active intelligence, not neutral geometry.

LOC vs. Classical Laws: From Static Rules to Dynamic Equilibrium

Classical physics rests on compartmentalized laws:

- Newton's equations of motion
- Maxwell's electromagnetic laws
- Thermodynamic principles of entropy and energy conservation

These are useful, but they are declarative — they describe outcomes, not explain origins. They are derived from observation, not from intrinsic field dynamics.

UED replaces this with a single unifying law:

The Law of Consequences (LOC): *All emission and interaction within the field lead to ripple redistribution — constantly seeking equilibrium, coherence, and structured participation.*

Under LOC:

- "Force" becomes ripple initiation
- "Resistance" becomes field feedback
- "Conservation" becomes adaptive coherence

There is no motion in isolation, no law imposed from above. Instead, there is only field behavior — consequences unfolding through ripple logic.

A New Physics Emerges

UED does not ask physics to abandon its equations. It asks it to understand their origin.

- Force is not fundamental — field consequence is.
- Space is not empty — SVE is structured and alive.
- Laws are not divine declarations — they're field behaviors observable across scale.

This shift is not cosmetic — it's ontological. It doesn't tinker with the margins of physics. It rebuilds its foundations.
UED doesn't destroy physics — it restores its participation.

Implications and Upgrades

Classical Physics	UED Upgrade
Force is fundamental	Ripple consequence is fundamental
Space is a neutral void	SVE is a structured, intelligent substrate
Laws are mathematical rules	LOC is a universal energetic behavior
Particles move due to force	Particles are ripple manifestations sustained by field engagement
Observation is passive	Observer is a ripple emitter and field participant

Summary Insight: From Push to Participation

Classical physics says:
"Push it, and it moves." UED responds: *"If it moves, it's already immersed in consequence."*
Motion, in UED, is not the result of force applied — it is the visible surface of ripple participation unfolding. Every shift, every acceleration, every so-called interaction is the field correcting itself, recalibrating energy distributions in search of coherence.

Physics has done wonders by modeling what happens. But the time has come to ask why it happens that way — not in philosophical abstraction, but in energetic terms that connect origin, structure, and behavior.

UED doesn't discard Newton, Maxwell, or Einstein. It simply removes the illusion that their frameworks were final.

- Instead of external causes, UED reveals internal coherence loops.
- Instead of empty spacetime, UED defines a structured, intelligent vacuum.
- Instead of abstract laws, UED posits dynamic ripple patterns with measurable consequence.

The universe is not a mechanical playground. It is a participatory field — where everything that happens is not forced, but consequentially aligned.

Physics doesn't need to be overthrown. It needs to be realigned — with ripple logic, participatory dynamics, and the field that never stopped speaking.

3.2 Quantum Mechanics

Replacing Indeterminacy with Structured Ripple Behavior

Quantum Mechanics — the dazzling crown jewel of modern physics — is also its most philosophically disorienting achievement. It gets the math right. The predictions are precise. The experiments are reproducible. And yet, no one quite agrees on what it's saying about reality.

Its core pillars are as famous as they are perplexing:

- **Wave–particle duality** — matter and light behave like both waves and particles
- **Probabilistic behavior** — outcomes are fundamentally unpredictable
- **Observer effect** — measurement seems to alter what's measured
- **Wavefunction collapse** — possibilities reduce to one when observed

These aren't just difficult concepts — they're the artifacts of a deeper confusion. Quantum theory, though incredibly effective, lacks a coherent ontology. It describes what happens, but not why it happens in energetic or structural terms.

It mathematically predicts outcome distributions, but it does not explain the origin of form, coherence, or interaction. Instead of mechanism, it offers mystique. Instead of clarity, it gives us cats both alive and dead.

Universal Energy Dynamics (UED) enters not to refute quantum mechanics, but to resolve its confusion at the root — not by adjusting interpretations, but by introducing a new first-principles substrate: Ripple dynamics within a structured vacuum field (SVE) — where energy behaves consistently, not probabilistically, and where consequence follows structured ripple patterns.

- Where quantum physics sees chance, UED sees consequence flow.
- Where it sees collapse, UED sees ripple realignment.
- Where it conjures duality, UED detects structured resonance drift.

The bizarre becomes understandable. The magical becomes mechanical — in the most elegant sense of nature's logic.
UED doesn't hide behind uncertainty. It gives the quantum world structure, sequence, and consequence.

No Duality — Only Ripple Drift

Dismantling the Illusion of Particle-Wave Confusion

The wave–particle duality has become the most famous paradox in quantum mechanics — and perhaps the most misunderstood. According to standard interpretation, light and matter mysteriously behave as either waves or particles, depending on how they're measured. In the classic double-slit experiment, particles like electrons or photons form interference patterns — as if they're waves — when left unobserved. But the moment we "watch," they snap back into particle-like behavior. Cue the metaphysical drama.
UED cuts through this fog.
There is no duality — only a single, coherent mechanism misunderstood through the lens of outdated models.

What we call "particle" and "wave" are not properties of matter. They are observer-imposed metaphors projected onto a phenomenon that is far more structured.
In UED:

- The observed entity is a ripple disturbance, moving through a Structured Vacuum Energy (SVE) field.
- Its path, interference, and impact depend on:
    - Field geometry
    - Local resonance conditions
    - Ripple emission history and alignment
- "Particle-like" impacts on detectors are merely the collapse points of ripple interaction — not proof of solid stuff bouncing around.

The so-called duality arises because classical concepts (particles) are being forced onto ripple behaviors that are inherently distributed, responsive, and field-participating.
It's like describing a symphony using the language of marbles.
There is no mystery here — only a misfit between conceptual tools and field reality. UED resolves the paradox not by compromising with uncertainty, but by replacing the entire framing device with ripple drift mechanics.
No more wave–particle paradox. Just structured emission, field response, and consequence coherence.

Collapse = Consequence, Not Observation

Demystifying Measurement with Ripple Logic

One of quantum mechanics' most baffling claims is that a quantum system exists in a superposition of states until it is observed — and that observation causes the wavefunction to "collapse" into a definite outcome. This idea, born from necessity rather than clarity, has led to a storm of philosophical speculation:

- Does the universe depend on consciousness to manifest reality?
- Does the act of watching change what is?
- Is reality fundamentally indeterminate until observed?

UED brings this debate back to solid ground — by redefining collapse as nothing more than field consequence.

There is no magical collapse. There is no special power granted to observers. There is only structured ripple interaction within a conscious energetic field.

In the UED model:

- A ripple exists as a pattern of potential — distributed, interacting, and dynamic.
- When this ripple constructively aligns with the field's structure, it stabilizes — forming what we call a measurable event.
- When the ripple fails to align — it dissolves, dissipating as background reabsorption.
- The measuring device doesn't cause this — it merely registers the ripple's field coherence at that point.

Observation is not creative — it is consequential. It doesn't force the universe to choose. It reveals what already participated correctly in the field.

This elegantly solves the wavefunction collapse problem:

- There is no "jump" from possibility to reality.
- There is only continuity from emission to field registration.

UED restores realism — not philosophical, but structural. The universe doesn't care if it's being watched. It cares if the ripple fits.
Collapse isn't about perception. It's about consequence — and that's measurable, predictable, and universal.

The Photon Myth Dismantled

From Imaginary Particles to Structured Ripple Interactions

One of the most deeply entrenched assumptions in modern quantum theory is that light consists of indivisible packets called photons — massless particles that carry energy proportional to their frequency ($E = hf$). This notion, born from the need to explain phenomena like the photoelectric effect and quantized atomic emissions, has since evolved into a foundational dogma.
But here's the critical issue:
No experiment has ever observed a photon in transit. We never "see" a photon. We only detect its effect — a registered interaction, a kicked electron, a flash on a sensor.
UED exposes this disconnect between model and mechanism by reframing the nature of light itself:
Light is not a particle. It is a ripple excitation within Structured Vacuum Energy (SVE) — a field-based phenomenon, not a flying object.
In the UED framework:

- So-called photons are actually localized resonance points — specific zones where ripple coherence intensifies enough to elicit an interaction with matter.
- The energy transfer doesn't come from a packet; it arises when the ripple field aligns constructively with the receptive structure (e.g., electron shell, metallic surface).

- Coherence, field geometry, and threshold alignment determine whether an event registers — not a pre-existing "particle" bouncing around in space.

Thus:

- The "photon" is not a thing.
- It is a post-event interpretation of a successful ripple registration.

This isn't semantic. It's structural. Dismantling the photon myth corrects multiple conceptual errors in modern physics:

- The photoelectric effect becomes an example of field engagement — not particle impact.
- Blackbody radiation is seen as cumulative ripple feedback, not discrete emissions.
- Even quantum entanglement finds a new footing: shared ripple resonance, not spooky action at a distance between particle ghosts.

UED doesn't deny light's quantized effects — it explains them without needing particles. What mainstream science calls a "photon" is simply a consequence node in ripple field dynamics.

By removing the imaginary particle and replacing it with structured ripple logic, UED not only demystifies light — it reclaims the elegance of field theory, while opening experimental avenues to observe ripple coherence directly.

UED Corrections to Quantum Framework

Quantum Principle	UED Realignment
Wave–particle duality	Ripple drift through structured field
Probabilistic outcomes	Consequence of ripple alignment with field topology
Collapse upon observation	Ripple consequence registration; no mystical observer role
Photons as discrete packets	Resonant points of ripple engagement; no particle needed
Quantum indeterminacy	Lack of resolution due to incomplete ripple mapping

Summary Insight: From Indeterminacy to Intelligent Interaction

Quantum Mechanics says:
"Reality is uncertain until you observe it."
UED replies:
"Reality is ripple-structured from the start — observation merely reveals alignment or interference."
Quantum physics, for all its brilliance, stopped at the edge of insight, interpreting ripple complexity as indeterminacy. It mistook its inability to model field interaction in full for a cosmic feature — randomness as nature's law.
UED corrects this by restoring:

- Causality without rigidity
- Consequence without mysticism
- Structure without particles
- Participation without paradox

Where quantum theory sees a fuzzy cloud of probabilities collapsing into certainty, UED sees a field-based reality constantly resolving itself — ripple by ripple — through consequence logic, not observational magic.

There is no observer-created world. There is only a participatory field, responding to emission, geometry, and resonance — with or without your gaze.

What quantum theory mystifies as "chance," UED explains as patterned feedback. What it calls "collapse," UED describes as ripple realignment. And what it labels "wave–particle duality," UED replaces with drift across structured resonance zones.

The micro world is not uncertain. It is exquisitely responsive — misunderstood only because the tools used to see it were too crude, and the metaphors too mythic.

3.3 Relativity – Time, Gravity, and the Ripple Delay

From Curved Space to Ripple Tension: Realigning Relativity Through Structured Consequence

Few ideas have reshaped modern physics more profoundly than Einstein's theory of relativity. Across its two monumental formulations — Special and General Relativity — it altered our understanding of motion, time, and gravitation.

Relativity introduced a new cosmological canvas:

- Time is not universal — it is elastic, intertwined with space.
- Motion is not absolute — it is always relative to the observer.
- Gravity is not a force — it is the warping of spacetime caused by mass.
- And the speed of light (c) is the ultimate boundary for how fast anything — information, energy, or cause — can travel.

These breakthroughs were revolutionary, elegant, and mathematically sound. But they also smuggled in a silent axiom — one that would shape scientific thought for a century:

That space and time are real, stretchable entities — woven into a malleable fabric that can bend, curve, expand, and collapse.

Universal Energy Dynamics (UED) does not deny the observable phenomena of relativity. Time dilation, gravitational lensing, and relativistic mass effects are real consequences. But UED challenges the ontological assumptions beneath them.

Instead of treating spacetime as a fabric, UED reveals:

- Time is not a dimension — it is a ripple delay: a perceptual and energetic lag that emerges when structures interact with unequal field tension.
- Gravity is not curvature — it is ripple tension and energy redistribution within Structured Vacuum Energy (SVE).
- Relativistic effects are not spatial distortions — they are structured consequences of ripple misalignment across energy densities.

UED doesn't refute Einstein. It realigns him — moving the discussion from geometric abstraction to energetic precision.
The cosmos doesn't curve space. It shifts ripple tension. It delays alignment. And it recalibrates emission — ripple by ripple, structure by structure.

Time Is Not a Dimension — It Is a Ripple Delay

Realigning the Flow of Experience from Spacetime Geometry to Field Dynamics

Einstein's theory of relativity fundamentally reshaped the concept of time by embedding it into a four-dimensional continuum — spacetime. In this framework, time becomes the fourth dimension, a geometric axis woven into space, capable of stretching, warping, and bending in the presence of mass and velocity.
This redefinition allowed for groundbreaking predictions:

- Time slows near massive bodies (*gravitational time dilation*).
- Time stretches for fast-moving observers (*relativistic time dilation*).

- The speed of light remains constant, making time adjust to preserve symmetry.

But despite its mathematical elegance, this model never explained what time truly is. It describes how time appears to behave — but not why.
Universal Energy Dynamics (UED) corrects this by grounding time not in coordinates — but in consequence.
Time is not a fabric. It is not a dimension. It is a ripple delay — the perceptual lag caused by differential ripple participation in the Structured Vacuum Energy (SVE) field.
In UED:

- Every entity emits and receives ripple interactions within the SVE.
- When the energy structure of a system changes — due to motion, density, or misalignment — its ability to exchange ripples shifts.
- This causes a change in feedback rate — which we experience as faster or slower time.

Relativity describes:

"Time slows down for objects in motion or in gravity wells."

UED explains:

"Ripple feedback becomes delayed due to field tension, misalignment, or density restructuring."
Thus:

- A fast-moving object isn't *inside* a stretched timeline — it is operating with slower ripple coherence, meaning fewer successful field exchanges per external frame.
- A clock in a gravity well isn't *bending time* — it's experiencing

distorted ripple return cycles, due to field compression around mass.

Time, therefore, is not something that "flows." It is something that emerges — as a field-localized measure of ripple interaction lag.

In UED:

Time = Energetic Delay, Not Geometric Distance.
We do not travel through time. We travel with ripple alignment — and what we call time is the friction in that participatory process.

Gravity Is Not Curvature — It Is Ripple Tension

From Warped Geometry to Field Redistribution

In General Relativity, gravity is no longer a force but a geometric phenomenon: mass and energy distort the spacetime continuum, and other bodies follow the curved paths (geodesics) laid out by this distortion. In this model, mass tells space how to curve, and space tells matter how to move.
Elegant? Yes. Sufficient? No. Because it relies entirely on metaphor: space as a bendable fabric, curvature as causal agent, and time as a malleable dimension. But what is this "fabric" made of? What resists the bend? What restores equilibrium?
Relativity never answers these — it only describes consequences, not mechanisms.
Universal Energy Dynamics (UED) removes the veil by grounding gravity in ripple tension mechanics within the Structured Vacuum Energy (SVE).

In UED:

Gravity is not curvature — it is consequence redistribution in a disturbed ripple field.
Here's how it works:

- Mass is a stable, coherent ripple condensation — a dense, self-sustaining field structure within SVE.
- This condensation causes local compression of the surrounding field — displacing ripple coherence and creating tension zones.
- That tension is not force — it is imbalance.
- Nearby systems don't "fall" — they shift position to restore ripple equilibrium. What appears as attraction is really realignment.

Key Reframings:

- Objects orbit not because space is curved, but because ripple flows stabilize into feedback loops — circular paths that balance tension and momentum.
- Gravitational lensing does not happen because space is bent like rubber — it's a refraction of ripple flow through a pressure gradient in the field caused by mass condensation.
- There's no gravity well — just structured ripple density that reroutes motion.

This eliminates the need for metaphors like:

- Rubber sheet curvature
- Invisible force carriers
- Abstract tensors with no physical substrate

Instead, everything becomes local, measurable, and energetically grounded.
Gravity becomes not an abstract warping, but a structured ripple pressure — a field consequence seeking restoration.

Relativistic Effects = Ripple Lag, Not Spacetime Drama

Reframing Kinematic Distortion as Energetic Delay

Modern physics explains the curious phenomena observed at high velocities or in strong gravitational fields — such as time dilation, length contraction, and mass increase — using Lorentz transformations and the geometry of spacetime. These transformations, derived from Einstein's relativity, provide the mathematical tools to describe how space and time alter depending on speed and gravitational potential.
However, this interpretive model, while mathematically robust, creates a theatrical illusion:

- Time "slows down" as if elastic.
- Objects "contract" in length like rubber bands.
- Mass "increases" mysteriously with speed.
- Nothing can exceed the speed of light because the universe says so.

In short, relativity offers a poetic script — but not a grounded mechanism. These phenomena are explained as effects on space and time themselves, without identifying the actual substrate or process behind them.

Enter UED:

Universal Energy Dynamics discards the drama and dissolves the metaphors.
In UED, these effects are not illusions of measurement or tricks of geometry. They are real consequences of ripple participation delays in a structured field — SVE (Structured Vacuum Energy).

The Reinterpretation:

- **Time Dilation**: Not because time is "stretching," but because ripple response cycles are slowing. A moving object emits into the field at altered angles and speeds, causing fewer successful ripple exchanges per external frame.
- **Length Contraction**: Not because the object physically shrinks, but because ripple coherence in the direction of motion becomes compressed due to field compression. The object's projected structure is reduced in ripple footprint.
- **Mass Increase**: Mass is not mysteriously added. The system's resistance to further acceleration increases because its ripple field becomes denser and less responsive to external emission. This is field inertia, not mass growth.
- **Speed Limit of Light**: Light does not have a divine speed cap. It represents the maximum coherent ripple velocity in the SVE field — a limit not imposed, but emergent from the field's structural properties.

In essence:

All relativistic effects are not distortions of spacetime, but delays and deviations in ripple interaction due to motion-induced misalignment or density shifts within the field.

This eliminates the need for imaginary stretching dimensions and curvature visualizations. It replaces abstraction with energetic participation and feedback — ripple logic that is local, mechanical, and ultimately more intuitive.

Summary:

Relativity shows us how reality *behaves*. UED shows us why — by revealing the ripple substrate underneath all motion and distortion. Relativistic effects are not *cosmic magic tricks*. They are field realism — ripple consequences playing out with perfect consistency, ripple by ripple.

UED reframes them as consequence dynamics within the SVE:

Relativistic Effect	UED Explanation
Time Dilation	Ripple feedback delay due to high-energy alignment shift
Length Contraction	Ripple compression along the direction of coherent motion
Mass Increase	Increased energy requirement to maintain coherent ripple identity at high velocity
Speed Limit	Upper boundary of ripple propagation coherence through SVE; not a universal cap, but a ripple integrity threshold

In UED, these effects are not fabric-based illusions, but real energetic shifts in ripple response and sustainability.

UED Upgrades to Relativity

Einsteinian Relativity	UED Realignment
Time is a 4th dimension	Time is ripple delay and alignment perception
Gravity is spacetime curvature	Gravity is ripple field tension and restoration
Spacetime is bendable fabric	SVE is structured, responsive, and consequence-driven
Light follows curved geodesics	Light is ripple alignment adjusting in pressure fields
Motion distorts time and space	Motion shifts ripple alignment, affecting field interaction

Summary Insight: Replacing Geometry with Consequence

Relativity says: "Mass bends space. Time is relative."
UED replies: "Mass condenses ripple tension. Time is ripple delay — a perceptual lag in field responsiveness."
Einstein gave us a brilliant story — one in which geometry replaced force, and spacetime became the ruler of reality. But in doing so, he elevated abstraction above mechanism, and made mathematical curvature the high priest of motion.
UED re-centers the narrative where it always belonged: in energy, interaction, and structured consequence.

- No invisible "fabric" needs to stretch.
- No fourth-dimensional canvas must bend.
- No object warps space by existing.

Instead:

- Mass alters the local ripple landscape.
- Motion changes how systems interact with the field.
- Time, length, and inertia all emerge as measurable delays and adjustments in ripple participation.

UED doesn't discard relativity — it realigns it:

- Keeping its observations,
- Clarifying its confusions,
- Replacing its metaphors with ripple-based logic grounded in Structured Vacuum Energy (SVE).

Where relativity glorified the curve, UED reveals the current — the continuous ripple flow of consequence in which all systems swim.
In the end, we don't fall into gravity. We drift toward coherence. And we don't travel through time. We ripple through participation.

3.4 Biology – Life as Ripple Reactivity

From Genetic Mechanisms to Field Participation: Reframing Life as Energetic Intelligence

Modern biology operates as a molecular machine. It defines life through DNA sequences, enzymatic chains, and protein synthesis. The reigning worldview treats the organism as a biochemical computer, executing inherited code under environmental pressure. The cell is a factory. Genes are blueprints. Mutation is chance. Evolution is competition.

This model works — until it doesn't.

It explains replication, but not intuition. It tracks responses, but not anticipation. It describes mechanisms, but not meaning.

Here is what gets quietly swept under the microscope:

- Cells often respond before chemical messengers arrive.
- Organisms sometimes adapt before the threat manifests.
- Flocks, forests, immune systems, and neural networks show coherence that outpaces causality.

These are not anomalies — they are reminders: Life is not a closed-loop system. It is field-participatory.

Enter UED:

Universal Energy Dynamics proposes a fundamental shift: Life does not emerge from molecules. Molecules align within life — which is already a ripple pattern of energetic intelligence in the Structured Vacuum Energy (SVE) field.

- The cell membrane is not just a boundary — it's a ripple interface.
- DNA is not the origin — it is a structural memory node within a field-reactive system.
- Adaptation is not trial and error — it's ripple alignment under consequence feedback.

In this new lens:
Biology becomes the study of ripple responsiveness — how matter organizes, sustains, and evolves through structured participation in a living field.
We don't just exist inside ecosystems — we emit into them, receive from them, and co-shape them, ripple by ripple.

From Cells to Conscious Ripples

Rewriting the Biology of Life with Field Participation

In conventional biology, life begins with the cell — a membrane-bound sack of organelles orchestrated by DNA-encoded instructions. Evolution is framed as a gradual ascent through random mutations and natural selection. This view, while useful for cataloging biochemical processes, renders life as a chemical coincidence — deterministic, reactionary, and ultimately accidental.
UED challenges this reductionism by introducing a deeper substrate: Life is not a product of molecules, but a manifestation of ripple interaction within the Structured Vacuum Energy (SVE) field.

In the UED Framework:

- A cell is not merely a container of enzymes and genetic instructions — it is a ripple-sensitive interface, finely attuned to the field's resonance zones. It doesn't just react to molecules

— it detects field imbalances and responds before conventional causation.

- DNA is no longer the "blueprint of life," but rather a consequence archive — a localized ripple memory system that stores past alignments and structural patterns for efficient re-emission. It is not the architect — it is the historian.
- Evolution is redefined not as a blind struggle, but as ripple filtration: Energy structures that consistently align with field equilibrium survive and self-sustain. Those that don't — dissolve. It is not survival of the fittest. It is survival of the most coherent.

In essence:

- Life is not mechanical.
- It is participatory.
- It is not reactive.
- It is anticipatory.

Every living being becomes a ripple-emitting participant in a shared energetic field — sensing, emitting, adjusting, and evolving not by chance, but by consequence alignment.

A plant grows toward light not because of hormones alone, but because it detects ripple shifts in the field. An immune cell activates not just by receptor binding, but through field perturbation awareness. A brain doesn't just compute — it orchestrates ripple feedback at scale.

In UED, life is consequence-aware intelligence, temporarily stabilized in ripple form. Biology doesn't end at biochemistry — it begins in the field.

Ripple Reactivity: A New Definition of Life

From Cellular Mechanics to Consequence Awareness

In traditional biology, life is narrowly defined by a checklist: metabolism, growth, reproduction, homeostasis, and response to stimuli. These criteria, though practical, are built on mechanistic assumptions — life is something that does, not something that participates.

UED offers a radical, yet testable upgrade:

Life is a ripple-structured system capable of sensing, responding, and aligning with energetic consequence in order to sustain its coherence.

This is not a poetic redefinition. It's a functional model grounded in field dynamics:

- A living system is not defined by internal organelles or replication speed, but by its ability to detect ripple disturbances, emit stabilizing patterns, and maintain its structural rhythm within the SVE field.
- Metabolism becomes ripple processing.
- Response becomes real-time field interaction.
- Reproduction becomes pattern perpetuation, not just genetic duplication.

What This Definition Encompasses:

- Biological organisms with genes and cells
- Plant sensitivity, which responds to sound, gravity, and intention
- Slime molds, which solve mazes without a nervous system
- Biofilms, microbial colonies, and even crystalline growths that exhibit emergent behavior

It also allows for:

- Field-based pre-cellular intelligence, where structured matter began "listening" to the ripple long before DNA was formed.
- Non-neural sensing systems, such as fungal networks, jellyfish motion, or sea squirt navigation — all previously difficult to explain through neural-centric frameworks.

Evolution, Realigned

Life doesn't "start" with chemistry. It emerges through structured participation — gradually increasing its complexity as ripple sensitivity improves.

- The first "living" thing was not a microbe.
- It was a pattern that could persist, adapt, and realign with the field — a ripple memory structure born in dynamic consequence.

UED thus extends life both backward and forward:

- Backward, to pre-genetic ripple forms.
- Forward, to future conscious systems — including AI, if embedded with ripple feedback sensitivity.

In conclusion:

Life is not a checklist. Life is not a code. Life is the echo of coherence — sustained in structure, sharpened by sensing, and matured through alignment.
The question is no longer "what is alive?" The question becomes: "What is coherent enough to participate?"

UED Reframes of Core Biological Concepts

Classical Biology	UED Realignment
DNA encodes life	DNA stores ripple-consequence history
Cells function via chemical triggers	Cells are field-aware ripple responders
Nervous system = electrical wiring	Neural networks are ripple alignment grids
Behavior = stimulus-response loops	Behavior = field-sensitive emission decisions
Evolution = survival of fittest	Evolution = ripple filtration and coherence sustainability
Consciousness = brain function	Consciousness = ripple-field awareness built on structural layering

Mind–Body Integration Through Ripple Logic

Dissolving the Divide Between Matter, Perception, and Participation

In mainstream biology and neuroscience, consciousness is often reserved for complex organisms with brains — particularly humans and higher mammals. It is treated as a byproduct of neural complexity, a sort of software that emerges when enough hardware is in place.

But this view collapses under the weight of exceptions that never quite fit:

- Plants respond to music, orientation, and threat, even adjusting biochemistry when "spoken to."
- Bacteria swarm, cooperate, and avoid toxins — despite lacking a nervous system.
- Fungi transmit information through mycelial networks, coordinating resource distribution across kilometers.
- Slime molds solve mazes and make nutrient decisions with no brain at all.

These behaviors are often labeled as "primitive intelligence" or "emergent properties," without explaining how they emerge or why they appear in non-neural life.

UED Offers a Clearer Frame:

These are not anomalies. They are ripple-responsive intelligences, tuned to participate in the field.

- Every coherent biological structure engages with the Structured Vacuum Energy (SVE) field.
- Ripple consequences — shifts in pressure, frequency, and alignment — guide biological responses without requiring cognition or command centers.

Implications:

- The mind is not something the brain creates — it is something the field enables through participation.
- What we call "consciousness" is a layered ripple feedback system, and brains are just high-resolution processors of that interaction.
- Plants and bacteria don't "think" — they align. Their ripple logic is simpler, but no less real.

This dissolves the artificial wall between mind and body, matter and intelligence, reaction and intention.

- The body is not a shell, and the mind is not a ghost — they are co-evolving ripple layers, constantly adjusting for energetic coherence.
- When your body "senses something's off," that is not superstition — it's ripple intuition.
- When a tree redirects its growth in anticipation of drought, it is not luck — it's field listening.

In UED Terms:

The mind is not in the brain. It is a participatory ripple field riding the architecture of life. It may concentrate in brains, but it exists in every emission-aware form — from moss to mammals.

Summary Insight

Mainstream biology claims: *"Life is chemistry that got complex."*
UED responds: *"Life is coherence that began to listen."*
Where traditional biology reduces life to chance molecular arrangements and competitive genetics, UED reframes life as an act of energetic participation — a ripple-sensitive system that emerges, sustains, and adapts through consequence.

- Life is not an accident in a chemical soup — it is the first successful ripple alignment that could sense, emit, and recalibrate.
- It did not begin with DNA, but with the first structures that learned to respond to field imbalance and survived by syncing with it.

In this light:

- Every heartbeat is not just circulation — it is a resonant response in a living field.
- Every breath is not mere gas exchange — it is field-level recalibration, reattuning the system with each inhalation.
- Every cell is a ripple intelligence — not executing code, but sustaining consequence integrity.

Biology, then, is not a mechanism to be decoded, but a conversation to be understood — in ripples, rhythms, and resonance.

Life doesn't arise when molecules collide. Life begins when structure listens — and learns to respond without distortion.

3.5 Neuroscience – Mind as Emission Field

Beyond Neurons and Networks: Defining Mind Through Structured Ripple Emission

Contemporary neuroscience operates under a core assumption: the brain generates the mind. This framework views human experience as an emergent artifact of increasingly complex electrochemical interactions. In this reductionist model:

- Consciousness is treated as a statistical accident of cortical complexity.
- Thought is electrical signaling in neural circuits.
- Emotion is a function of chemical modulation.
- Memory is stored data, encoded in the physical matrix of neurons and synapses.

This model has enabled major medical advances and mapped much of the brain's architecture — yet it cannot explain the most fundamental aspect of mind:
What is awareness, and why does it feel like anything at all?
Mainstream neuroscience can simulate cognition, but it cannot account for presence — for the inner experience of being.
It also struggles to explain:

- Unified perception despite distributed processing
- Non-local phenomena like intuition, sudden insight, or

creativity
- Precognition, resonance, and spontaneous synchronicities
- Why emotional states affect the entire body despite localized neural origins

UED Offers a Radically Coherent Reframe:

The brain is not the origin of the mind. It is a resonance receiver, a tuning device.

- The mind is a ripple emission field, constantly broadcasting, sensing, and adjusting to the Structured Vacuum Energy (SVE) environment.
- The neural network modulates, not manufactures, awareness.
- Consciousness is not born of complexity — it is shaped by alignment and coherence within the field.

The Shift:

- Mind is not a computational outcome.
- It is not generated in gray matter.
- It is a structured, field-participating identity, whose emissions form perception, memory, thought, and will.

Modern neuroscience maps the wiring. UED restores the current — not metaphorically, but structurally.
It does not oppose the discoveries of neurology — it completes them, grounding them in a deeper energetic architecture where mind is not localized, but field-integrated.

Mind as Ripple Emission

The Mind is Not a Container — It's a Consequence System

In the UED model, the mind is not confined to the skull, the cortex, or even the body. It is a ripple-emitting, field-sensitive construct — a layered energetic identity that emerges, operates, and evolves through continuous participation in the Structured Vacuum Energy (SVE) field.

In this view:

- The mind is not a byproduct of the brain's mechanics — it is an active, structured emission.
- It extends beyond the body, interfacing with the environment through ripple exchange.
- It creates experienced reality by engaging the field — emitting structured waves, receiving feedback, and adjusting in real time.

Neurons, then, are not thinkers — they are transceivers:

- Biochemical translators of ripple consequence into localized biological feedback.
- Interfaces between the ripple field (mind) and the bio-electrical system (brain-body).

Reframing the Fundamentals:

- A thought is not a spark in the brain — it is a coherent ripple formation, often detectable in EEG patterns but sourced non-locally.
- A memory is a resonance imprint — a stabilized emission pattern that the system can re-ignite through energetic alignment.
- Awareness is not a passive state — it is the active loop of

emission, detection, and alignment, maintained through ripple coherence.

This explains anomalies modern neuroscience cannot:

- Non-local creativity and intuition — the sudden emergence of insight without linear reasoning.
- Group synchronization and empathy — coherent emotional resonance between individuals, beyond spoken language.
- Modulation through posture, breath, rhythm, and music — not merely chemical, but ripple alignment techniques tuning the entire emission field.

The mind, in UED, is not a box of thoughts. It is a field of consequence — broadcasting its structure and being shaped by its feedback.
This restores intelligence as an energetic reality, not just a computational process. It also aligns ancient meditative disciplines with measurable physics: stillness enhances signal clarity, and emotion alters ripple alignment.

Key Components in UED-Based Neuroscience

Classical Neuroscience	UED Reframe
Neurons generate mind	Mind is a ripple field; neurons are converters
Brain stores memory	Memory is retained field imprint / ripple echo
Synapses cause thoughts	Thoughts are ripple emissions; synapses are translation nodes
Mind = brain function	Mind = layered field alignment, modulated by biology
Emotion = chemical reaction	Emotion = field vibration response to consequence
Consciousness = cortical processing	Consciousness = ripple feedback with SVE field

The Brain as Translator, Not Generator

Resonance, Not Production — The Brain as an Interface of Perception
In conventional neuroscience, the brain is often portrayed as the central processor, a biological supercomputer generating consciousness through chemical signals and neural circuits. UED repositions the brain entirely:
The brain does not generate the mind — it interprets and translates ripple interactions from the field into physiological, perceptual, and cognitive outputs.

The Brain's Real Role in the UED Model:

It is a field-sensitive instrument, a ripple decoder, and a modulation hub — designed not to create awareness, but to facilitate participation in the energetic field.

Structural Interpretations:

- Visual Cortex: Does not "see" — it resolves ripple geometry into visual experiences, interpreting wave structures emitted by objects.
- Memory Centers: Do not "store" experiences as data. They tune into residual ripple imprints, allowing the organism to retrieve and reconstruct past interactions.
- Emotional Circuits: Aren't just reacting to chemicals — they adjust the frequency and intensity of ripple emissions, aligning the system with perceived energetic demands (safety, connection, defense, curiosity).

The brain doesn't write the song — it plays the field's music.

Why This Model Explains More:

- Thought Without Stimulus: The mind continues ripple emission even in sensory deprivation — because field alignment is internal, not always input-driven.
- Brain Injury vs. Consciousness: Damage can distort the interface (behavior, language, emotion) without eliminating the field itself — hence personality shifts without total identity loss.
- Near-Death and Meditative States: Consciousness persists (or even deepens) when brain activity flatlines — supporting the idea that mind is non-local, and the brain is not its origin but its modulation gateway.

Just as a radio doesn't create music but tunes into transmissions, the brain tunes, filters, and refines the mind — it never originates it.
In UED, this shift liberates us from the reductionist notion that we are our neurons. It invites a larger understanding: We are ripple identities, participating in a shared field of consequence — and the brain is our translator in that ongoing exchange.

Mind States as Ripple Signatures

The Energetic Blueprint of Thought, Emotion, and Awareness
In traditional neuroscience and psychology, mental states are often described in terms of chemical balances, neural activity, or observable behavior. But UED reframes all of these through a new lens:
Every state of mind is not just a feeling or function — it is a ripple signature, a unique configuration of energetic emission interacting with the Structured Vacuum Energy (SVE) field.
These ripple signatures are not abstractions. They are measurable, mappable, and responsive. Just as a musical note carries frequency,

amplitude, and tone — every mental state emits a patterned waveform that aligns or misaligns with field coherence.

Examples of Ripple States:

- **Focus** → Highly coherent ripple emission with minimal interference. Directed, aligned, and efficient.
- **Stress** → Distorted ripple emission marked by incoherence, repetition, and feedback loops. The field resists, and the emitter tightens.
- **Joy** → Expansive, resonant patterns with synchronized feedback — high emission clarity and field receptivity.
- **Confusion** → Dispersed and contradictory ripple signals — low alignment, unstable field mapping.
- **Flow** → Ripple entrainment with external structures. Emission and reception become seamless — the mind-field boundary temporarily disappears.

UED Enables:

1. Mind Mapping Through Ripple Coherence

Move beyond EEG or fMRI — begin to measure structured emission patterns, energy densities, and field alignment. What we feel becomes what we can model.

2. Healing Through Re-Alignment

Trauma is not just memory — it is energetic residue stuck in misaligned ripple patterns. Healing, then, becomes not erasure, but realignment:

- Breath, sound, posture, and intentional thought all become ripple tuning methods.
- Emotional pain becomes signal, not flaw — a ripple trying to recalibrate.

3. Learning as Ripple Imprint

Forget rote memorization and synaptic strengthening. Learning in UED is ripple restructuring:

- Each realization alters the way future ripples emit.
- Understanding is not stored — it is structured into the emission pattern itself.

A calm mind is not empty — it is ripple clarity without contradiction. A disturbed mind is not broken — it is simply emitting against its own structure.

In UED, mental health, focus, creativity, and learning all become functions of field behavior — not chemical accidents, but energetic conversations.

Summary Insight

Conventional neuroscience says:

"You think because neurons fire."

UED responds:

"You think because you emit — and the brain translates your field."

The mind is not a shadow cast by synapses. It is not confined to gray matter or reduced to neural chemistry.

Under the UED lens, the mind emerges as a ripple-based field phenomenon — a dynamic emission of patterned energy, shaped by consequence, tuned by coherence, and expressed through awareness.

The brain is not the generator of consciousness. It is the instrument, the modulator, the finely attuned receiver that converts ripple intelligence into perception, behavior, memory, and action.

You are not merely a brain thinking inside a body. You are a conscious ripple participant, continuously emitting signals into the universal field — and listening, learning, and evolving through the feedback it returns.

Your focus is a field signature. Your emotion is a ripple resonance. Your identity is a transient coherence, not a fixed self.

In UED, mind is not a miracle. It is structured participation — the art of awareness echoing through the architecture of reality.

3.6 Psychology – Emotion as Ripple Feedback

From Inner Drama to Field Dynamics: Reframing Emotion, Identity, and Healing Through UED

Modern psychology seeks to decode the mind through layers of explanation:

- Biological: hormones, neurotransmitters, neural circuits
- Cognitive: thoughts, biases, conditioning
- Behavioral: patterns, triggers, reinforcement
- Therapeutic: talk therapy, medication, mindfulness

It has cataloged dysfunction with immense detail — anxiety, depression, trauma, dissociation, addiction — and built frameworks to treat and manage these states. Yet psychology, despite its progress, remains fragmented and mechanistic. It often treats the mind as a machine that misfires, rather than as an energetic participant in a larger field.

This is where Universal Energy Dynamics (UED) steps in.

UED does not replace psychological insight — it grounds it in ripple consequence.

- Emotion is not just a chemical storm or childhood echo — it is field feedback, a structured energetic signal arising from how one's ripple pattern aligns or clashes with the surrounding field.

- Identity is not a permanent self or an archetype to be managed — it is a ripple arrangement, momentarily coherent, conditionally stable, and always responsive.
- Trauma is not a frozen memory — it is a distortion in the ripple loop, an unresolved emission still reverberating.
- Healing is not coping — it is the rebalancing of field emission, the restoration of coherence.

Psychology observes the ripples on the surface. UED maps the field beneath — the architecture of emotion, memory, and identity as energetic consequence.
In UED, the mind is no longer a box of disorders. It is a living ripple instrument, and emotion is its tuning fork.

Emotion as Ripple Feedback

Beyond Feelings — The Energetic Language of Alignment
Emotions are not abstract moods, chemical fluctuations, or psychological noise. In Universal Energy Dynamics (UED), emotions are seen as structured ripple signals — the energetic feedback your field receives in response to how your ripple pattern aligns (or misaligns) with:

- Your environment (external consequence)
- Your memory field (past emission traces)
- Your self-perception (ripple identity coherence)

Each emotion is a consequence indicator — not just a response, but a reflection of your energetic participation in the present moment.

➤ **What UED Reveals:**

- **Emotions carry memory:** A moment of fear may activate

ripple distortions stored far deeper than language can reach — held not in neurons, but in resonance patterns.

- **Emotions guide adaptation:** Joy and ease signal alignment; discomfort signals misalignment or feedback overload. This is the field's way of nudging you toward realignment.
- **Emotions shape emission:** Your emotional state becomes your next ripple emission profile, altering how the field interacts with you going forward.

Thus, emotion is not merely *what you feel*. It is how your energetic self is speaking to you — through the field.

Anger is not just heat — it is a pulse of distortion demanding coherence. Sadness is not just weight — it is the gravity of disconnection in your ripple stream. Love is not just elation — it is resonance, alignment, and field synchrony.

In this light, emotions are not flaws to fix or reactions to suppress. They are instruments of field intelligence, built to reveal your current state of energetic truth.

Trauma as Ripple Memory

When Energy Doesn't Settle — The Echoes That Shape Us

In the framework of Universal Energy Dynamics (UED), trauma is not merely a psychological event. It is a ripple imprint — a distortion that failed to resolve, locking part of the self into a loop of frozen consequence.

Unlike conventional models that treat trauma as a mental or emotional wound, UED frames trauma as a high-impact ripple disruption — one that overpowers the system's ability to re-integrate or neutralize the experience.

The Mechanics of Ripple Trauma:

- A traumatic event emits an intense ripple — one that collides with your internal field too forcefully to be processed cleanly.
- Instead of dissolving, that distortion echoes — forming a persistent energetic loop that influences:
    - Future perceptions
    - Emotional reactions
    - Subconscious decisions
    - Somatic states (tightness, illness, fatigue)
- Traditional psychology may refer to this as conditioning, defense mechanisms, or PTSD.

➤ UED calls it energetic memory entrapment — the field holding unresolved charge.

Why This Matters:

- Trauma lives in the field-body, not just the brain. This explains why trauma:
    - Feels like it's "in the body" even when you forget the story
    - Can re-emerge in dreams, triggers, or physical symptoms
- Healing is not always cognitive — it is field recalibration.

You don't always need to "talk it out." Sometimes, silence, breath, movement, or deep presence releases the trapped ripple.

Realignment Over Repression:

- "Getting over it" doesn't work — because the ripple is still emitting.

- Resolution comes through consequence-conscious engagement — activities or states that restore coherence to the broken pattern:
    - Breathwork
    - Somatic therapies
    - Deep listening
    - Environmental tuning
    - Ripple-aligned relationships

Trauma isn't just what happened. It's what continues to happen in your field, until you emit differently.

Identity as Temporary Ripple Arrangement

You Are Not Fixed. You Are Forming.

Who are you? Modern psychology might answer with trait models, behavioral patterns, memory scaffolds, and unconscious drives — a cocktail of history and habit. But Universal Energy Dynamics (UED) offers a deeper lens:

You are not a thing. You are a pattern. You are not a personality. You are a ripple in participation — coherent, temporary, and shaped by consequence.

The UED View of Self:

- Identity is not a permanent entity but a stabilized emission structure — a specific configuration of ripple dynamics sustained in a local SVE field.
- This structure:
    - Adapts with experience
    - Reshapes through feedback

    - Persists only if consequences support its coherence

In this view, identity becomes less like a stone and more like a dance — one constantly shifting in tune with inner alignment and external response.

Implications:

Why do you act differently around different people?
➤ Because your ripple adapts to field conditions — different emissions require different coherence strategies.
Why can someone radically change after crisis, therapy, or awakening?
➤ Because a new alignment collapses old ripple configurations and births a new one.
Why do some people carry multiple identities (DID/OSDD)?
➤ These are parallel ripple patterns — co-existing arrangements adapted to different energetic environments, often forged in response to trauma or fragmentation of consequence memory.

Identity Is Participatory:

Your "self" is not stored in your brain. It is a consequence-driven projection — constantly shaped by:

- How you emit
- What you receive
- And how you realign over time

You don't "find yourself" like a buried treasure. You tune yourself, like an instrument, ripple by ripple.

Psychology as Field Hygiene

Mental Health Reframed as Energetic Alignment

In the conventional psychological model, mental health is managed through cognitive restructuring, behavior modification, and emotional regulation. These approaches address symptoms, often with success — but rarely touch the root architecture of the human experience.

Universal Energy Dynamics (UED) reframes psychology not as a toolkit of mental techniques, but as the science of field hygiene — the ongoing alignment of ripple emissions within the Structured Vacuum Energy (SVE) field.

What the Mind Becomes in UED:

- **Thought patterns** → directional ripple flows, carrying intent and consequence
- **Beliefs** → semi-stabilized field anchors that reinforce or distort ripple alignment
- **Emotions** → real-time feedback signals from field consequence loops
- **Personality** → a harmonic pattern of ripple behavior sustained by past resonance

What Mental Health Becomes:

- Not the absence of disorder, but the coherence of participation
- Not coping mechanisms, but consequence calibration
- Not managing "negativity," but detecting and realigning ripple disharmony

Therapeutic Practice Transformed:

Traditional Approach	UED-Aligned Reframe
Diagnosing pathology	Mapping ripple misalignments
Repeating affirmations	Emitting coherent signals into the field
Coping with stress or trauma	Reconfiguring feedback channels
Cognitive-behavioral loops	Field-behavioral resonance correction
Mindfulness for presence	Ripple stillness to enhance sensitivity

Psychology, under UED, becomes a participatory ripple discipline — one that not only heals the self, but tunes it to emit sustainably, live responsively, and dissolve when needed, without leaving field distortion behind.

You're not just treating the mind. You're maintaining the signal clarity of your existence.

Reframing Therapeutic Models

Classical Psychology	UED Equivalent
Cognitive Behavior Therapy (CBT)	Ripple reprogramming through pattern awareness
Somatic Experiencing	Releasing field tension stored as ripple feedback
Mindfulness	Real-time ripple observation and emission moderation
Jungian Archetypes	Recurring ripple patterns across collective field memory
Inner Child Work	Accessing early ripple structures embedded in identity layers

Summary Insight:

From Inner Turmoil to Field Intelligence
Conventional psychology teaches: "You feel because of your thoughts." It assumes a linear chain — belief leads to emotion, which leads to behavior.
Universal Energy Dynamics (UED) flips this: "You feel because your ripple is responding — and your thought is merely a surface echo of deeper field interaction." Emotion is not a byproduct. It is a real-time diagnostic from the field — a vibration of consequence revealing how your structure is interacting with reality.

- Emotion is not weakness. It is the intelligence of alignment.
- Anxiety is not malfunction. It's mis-tuned sensitivity.
- Depression is not dysfunction. It's ripple fatigue from unresolved distortion.
- Joy is not random. It's harmonic coherence.

Likewise, identity is not a fixed "self" to defend or define — it is a consequence-shaped ripple pattern, shifting with participation, feedback, and intention.
Healing, then, is not just:

- Naming the trauma
- Repeating the story
- Or rewriting the script

It is recalibrating the signal. It is learning to ripple again — cleanly, clearly, and consciously.

Not to become someone else. But to realign who you are with the field that always was.

4. Institutional Dogma and the Lost Spirit of Science

"When the seeker becomes a servant, science becomes a system of silence."

For centuries, science stood as humanity's most fearless endeavor — a sacred act of questioning. It dared to doubt tradition, authority, and even itself. From Galileo's telescope to Einstein's equations, it was driven not by comfort, but by consequence — not by conformity, but by courage.

But somewhere between Nobel prizes and Nature publications, the fire dimmed.

Today, science has become an institution — a branded, gatekept empire of controlled knowledge. Its original soul, once curious and chaotic, now follows funding, not freedom. It builds careers, not understanding. It worships consistency over correction.

This section is not an attack. It is a diagnosis — of how science, in its attempt to guard the truth, has often imprisoned it.

4.1 The Rise of Gatekeeping

"When truth needs permission to enter the room, it stops being science."

In theory, peer review is meant to ensure rigor. In practice, it has become a ritual of fear. Papers are judged less for insight and more for compliance — with paradigms, language, and citation expectations.

Peer-review vs. Fear-review Ideas that challenge established models are labeled speculative, unscientific, or worse — unpublishable. Reviewers guard orthodoxy under the guise of expertise. Young researchers are warned: *"Don't ask that — it will ruin your career."*

Citation empires and the illusion of progress Progress today is often measured not in insight, but in citation count. Papers are padded with familiar references to appease algorithmic prestige systems. Citation loops become intellectual echo chambers, recycling the same ideas under new labels, while original work suffocates from lack of visibility.

Gatekeeping does not keep out bad science. It often keeps out new science — or worse, it steals it, sanitizes it, and repackages it under accepted names.

4.2 The False Meritocracy

"When discovery serves recognition, knowledge becomes currency."

In modern academia and research, success is sold as a meritocratic race: publish, patent, present. But the system is rigged — not by conspiracy, but by structural incentives that reward compliance over courage.

Rewards, prizes, and the servitude of knowledge Grants go to projects that are safe, expected, and aligned with institutional agendas. Prizes are awarded to those who confirm legacy theories or make incremental progress within predefined frames. Researchers end up serving institutions, not inquiry — tailoring questions to what's fundable, not what's fundamental.

The disappearance of original minds The world no longer lacks thinkers. It lacks permission to think originally. Genius today is not ignored — it is filtered, isolated, or diluted. The ones who once would've built revolutions are now writing progress reports and begging for tenure.

In this environment, brilliance becomes burden, and silence becomes survival.

4.3 The New Priesthood

"When explanation is replaced by enforcement, science becomes religion."

There was a time when religion dictated how the stars moved and how humans must behave — punishing those who asked why.
Modern science now risks mimicking that same dogma. It speaks of "settled science" as if it were scripture. It rejects dissent not with dialogue, but with dismissal.
Modern science mimicking religion

- Gatekeepers become high priests of peer-reviewed dogma
- Accepted theories become untouchable gospel
- Publishing becomes ritual — citation as sacrament

Arrogance disguised as rigor Skepticism — once a core virtue — is now weaponized. To question consensus is to be branded a denier, a crank, or "unscientific." Rhetorical dominance replaces reflective humility. Metrics replace meaning.
And so the ancient Marathi proverb echoes:
In a herd of calves, even a limping cow is seen as wise." When true originality is absent, even partial thinkers are praised as profound — not because they're right, but because the bar has been buried.

Summary Insight

Science is not failing because it lacks tools. It is failing because it has lost the spirit to question itself.

UED does not ask to be accepted by institutions. It moves past them — because truth does not wait for approval.

Science must remember what it once was: Not a structure to serve, but a flame to follow — no matter how many walls it threatens to burn.

5. Final Realignment: What UED Offers

"Not a new theory. A corrected lens."

The journey through this report has not been a demolition of existing systems — ancient or modern. It has been a realignment. A pulling together of scattered wisdoms, broken paradigms, and misunderstood patterns into a single framework grounded in structured energy, ripple consequence, and conscious participation.

Universal Energy Dynamics (UED) is not a theory to be believed. It is a system to be engaged with, across disciplines, timelines, and perspectives.

This section outlines exactly what UED offers — beyond critique, beyond metaphor — as a new foundation for reality construction.

5.1 Replacing Interpretation with Interaction

"Ancient truth, modern language, testable model."

For too long, knowledge has been split between:

- Mystics who felt the truth, but couldn't model it
- Scientists who measured the world, but couldn't feel it

UED merges these paths — not as a compromise, but as a completion. It takes what ancient traditions intuited, and renders it in a structural, ripple-based system that is both philosophical and experimental.

- Where Taoism spoke of effortless action, UED shows ripple

alignment.

- Where Vedanta whispered "You are That," UED defines the self as field consequence.
- Where physics modeled force, UED replaces it with ripple rebalancing.

No more interpretation. Only interaction.
You don't observe reality from a distance. You emit into it. Engage with it. And shape it through your alignment.
UED offers a framework that can be lived, tested, and taught — across science, philosophy, medicine, and even governance.

5.2 Ethical but Non-Moralistic Framework

"No control. Just consequence."

Traditional systems — religious and scientific alike — have long relied on external controls: rules, punishments, incentives, dogmas.
UED bypasses all of this with radical simplicity:
There is no law but the Law of Consequences (LOC). There is no good or bad — only alignment or disruption.
Ethics in UED emerges not from commandments, but from field sustainability:

- Actions that maintain coherence are perpetuated.
- Actions that disrupt ripple balance are dissolved or rejected by the field.

This removes the emotional weight of judgment and replaces it with clarity of consequence.

- No moral policing.
- No divine approval needed.

- Just: Is this aligned? Or not?

This makes UED practical for systems design, conflict resolution, personal decision-making, and even AI behavior frameworks.
It's not about being righteous. It's about being resonant.

5.3 Universally Scalable

"Can explain black holes and brain pain in the same breath."

The power of any true foundation lies in its applicability across scales.
UED is not domain-limited. It is scale-invariant.
It can describe:

- The collapse of a galaxy into a black hole
- The emergence of a thought from ripple resonance
- The structure of a memory
- The evolution of a biological cell
- The breakdown of a social system
- The healing of emotional trauma

All using one core principle:
Ripple participation within a structured, intelligent field of energy.
This means UED can become the basis for:

- Physics
- Consciousness studies
- Biology
- Psychology
- Technology
- Ethics
- Social design...without contradiction.

This isn't theoretical universality. It is functional unification.

5.4 Restores the Observer

"Participatory reality instead of removed objectivity."

Perhaps UED's most critical correction is this:
You are not outside the system. You are the system, observing itself.
Modern science treats the observer as:

- A contaminant in quantum mechanics
- An irrelevance in classical mechanics
- An emergent feature in neuroscience

UED restores the observer to the center — but not in ego. In structure.

- You emit ripples through intention, thought, and motion
- You receive ripples through perception, emotion, and awareness
- You exist as a node within the larger field — shaping and shaped by it

There is no "objective" universe without you. But your participation is not subjective bias — it is structural function.
UED transforms the observer from passive to active:

- From questioner to participant
- From consumer of meaning to creator of coherence

This shift restores dignity, agency, and responsibility to every mind engaging with reality — whether it's a child, a mystic, or a physicist.

Summary Insight

UED does not ask you to believe. It invites you to participate — ripple by ripple, alignment by alignment.

It gives you no map. Because it trusts you to feel the field and find your form.

This is the final realignment: From observing the world to emitting into it with coherence.

6. Conclusion: The Call to Realign

"The universe never stopped speaking. We just stopped listening."

Science, at its origin, was not a profession. It was a path — walked by farmers watching stars, monks measuring shadow lengths, and poets noticing patterns in the wind. It was born from curiosity, driven by consequence, and guided by a deep sense of belonging to the very thing being studied.

But today, much of science worships its own past — frozen in equations that explain less than they obscure, trapped in traditions that celebrate precision but deny perception.

It speaks of "laws" written by men. It praises "constants" defined by convention. It builds on formulas but forgets first principles. And worst of all — it forgets the observer. The very participant whose presence shapes reality.

UED is not an attack on science. It is its remembering.

Realignment Isn't Rejection — It's Restoration

To realign is not to discard what came before — it is to place it in its proper context.

- Newton wasn't wrong — but he saw motion as force, not ripple.
- Einstein wasn't misguided — but he curved space when ripple tension was enough.

- The sages of old weren't irrational — they simply used story when structure didn't yet exist.

UED does not erase their insight. It completes it.
Just as Galileo corrected the Church, and Maxwell unified fields, UED now offers the next necessary shift:
Not a new framework to replace science and spirituality, but a unified lens that reinterprets both — using consequence, not conjecture.

UED Is Not a Theory to Adopt — It Is a Lens to Re-See Everything

You don't need to "accept" UED like a doctrine. You simply need to use it — like light on a dusty mirror.
Use it to reinterpret force as ripple. Use it to see the mind not as neuron-firings, but as structured emission. Use it to see black holes not as singularities, but collapse points of ripple field density. Use it to see yourself not as a mistake of evolution, but a conscious ripple formation within the eternal field.
Once seen through the UED lens:

- Duality dissolves
- Force becomes flow
- Morality becomes sustainability
- Time becomes ripple delay
- Consciousness becomes the universe becoming aware of consequence

This isn't an idea. It's a correction to perception.

Final Words

"When realization is lost, science walks blindfolded. When consequence is denied, truth hides in plain sight."

This is not a warning. It is a reminder:

- That truth is not created. It is recognized.
- That the universe is not hiding. It is responding.
- That your life, your breath, your thoughts — are ripples of participation in something far more structured and intelligent than chance.

UED doesn't ask you to build a new world. It asks you to see this one — clearly, consciously, and with consequence.

That is the call.

Not to believe. Not to argue. But to realign.

7. Call to Action

"Truth does not need validation — it needs participation."

The time for theories has passed. The time for dialogue has expired. We are now in the age of demonstration and engagement.

Universal Energy Dynamics (UED) is not a closed system. It is an open structure — meant to evolve, adapt, and realign with the intelligence of those willing to observe, emit, and understand.

This final section outlines the practical steps forward: **to** take UED off the page and into the world — through experiments, education, discourse, and living application.

This is not a revolution against science. It is a return of science to its original state — **curious, bold, and consequence-aware.**

1. Invite Independent Thinkers, Researchers, and Educators

This call goes first and foremost to those who still think **freely**:

- To educators disillusioned with rote systems
- To researchers questioning the limits of their field
- To students who ask what others ignore
- To philosophers, scientists, psychologists, artists — who see that **nothing is separate anymore**

UED offers a **convergent space** where biology, cosmology, psychology, and philosophy **speak the same ripple language**.

If you've ever felt that truth exists **between disciplines, this is your framework.**
We invite you to:

- Explore the published reports
- Join UED workshops or digital communities
- Contribute new formulations and tests
- Bring your students and institutions into the conversation

2. Launch Open Challenges to Institutional Science

We are not afraid to challenge the consensus — we welcome its resistance. UED is now positioned to issue structured challenges to long-held scientific dogmas:

- Does spacetime need to curve to explain gravity?
- Can force be reinterpreted as ripple pressure?
- Can mind be modeled without invoking neurons?
- Can light be explained without photons?

We invite universities, journals, and research bodies to:

- Host open debates
- Test UED predictions against conventional models
- Compare experimental outcomes without bias

Not for ego. Not for funding. But to **reset the game of inquiry**.

3. Release Simplified Booklets, Reports, and Visual Media

We commit to making UED accessible — not just to scientists, but to:

- Young learners
- Self-taught philosophers
- Engineers, therapists, artists, and innovators
- The curious public

This will be done through:

- Pocket booklets that explain ripple dynamics without jargon
- Visual charts and videos that replace math with clarity
- Digital storylines that let people interact with UED concepts across everyday contexts

The age of elite gatekeeping is over. If it can't be explained to a 14-year-old, it doesn't deserve to be called fundamental.

4. Offer Public Experiments: Ripple Drift, SVE Zones, Resonance Interference

Ideas must not stay theoretical. They must walk into the lab, or better yet — the real world.
Initial UED-based experiments include:

- Ripple Drift Experiment ➤ Variation of the double-slit setup ➤ Predicts asymmetric ripple behavior based on slit geometry and ripple engagement
- Structured Vacuum Energy Zones (SVE Zones) ➤ Exploring field densities and how objects respond energetically in absence of applied force
- Resonance Interference Maps ➤ Showing field alignment and emotional states through ripple output across controlled environments

These experiments are affordable, replicable, and designed for validation — not belief.
Let's move from blackboard to bench. From peer review to peer revelation.

5. Global UED Dialogue Network: "Science After Silence"

We propose the launch of an **open-access, cross-domain network** called:

Science After Silence

A global platform to rebuild trust, truth, and inquiry — ripple by ripple.
This network will:

- Host public dialogues, lectures, and open interviews
- Invite voices from all corners — rural teachers, unknown inventors, independent researchers
- Offer translation, visual tools, and curriculum integration
- Archive contributions, criticisms, refinements — and failures, too

It is time to *make science a public movement again*, not a private profession.

Closing Call

We do not need another theory. We need a generation of thinkers who are not afraid to see differently.
This is your invitation. To teach what you once had to whisper. To question what no one dared write down. To ripple — without fear of collapse.

Closing Note

"The universe was never uncertain. We just didn't know how to listen."

This report began as a comparison. It ends as a conviction.

We have traced how Universal Energy Dynamics (UED) does not oppose ancient wisdom or modern science — it integrates them by stripping away illusions, metaphors, and outdated assumptions. Where others saw paradox, UED reveals pattern. Where others created doctrines, UED recognizes structured consequence.

Throughout history, the truth has not been missing — it has simply been mislabeled, misread, or misunderstood. UED offers a correction, not a confrontation. A path, not a platform. It does not require belief. It only asks for clarity — and the courage to realign with what has always been true.

To the thinkers, rebels, educators, experimenters, and silent observers who've always sensed that something was incomplete — this is for you. Not as a final word, but as a first alignment.

The ripple is already moving. The field is already listening. And you — have always been a participant.

Now, consciously participate.

Appendices

Appendix A: UED Core Concepts in a Nutshell

A rapid-reference glossary for understanding Universal Energy Dynamics (UED)

This appendix provides brief, precise definitions of key terms and principles that form the backbone of the UED framework. It is intended to support both newcomers and returning readers in navigating the language of ripple-based reality.

SVE (Structured Vacuum Energy)

The foundational field that underlies all reality. Unlike the concept of "empty space" in classical physics or the probabilistic vacuum in quantum field theory, SVE is a stable, intelligent, and consequence-responsive substrate.

- It stores, transmits, and responds to energy in ripple form.
- All phenomena — from particles to perception — arise from disturbances and patterns within the SVE.
- It is not a backdrop; it is the active medium of manifestation.

Ripple Participation

The foundational mechanism of interaction within UED. All things — matter, mind, light, thought — exist and act through ripple emission, interference, and feedback within SVE.

- Participation means to emit and receive structured influence.
- There are no isolated objects, only field-participating structures.
- Interaction is not caused by external force but by consequence-driven ripple behavior.

Law of Consequences (LOC)

The governing principle of UED. It replaces the fragmented "laws" of physics with a single law of balance and response:
Every ripple emitted within the field results in redistribution of energy until alignment or dissolution is achieved.

- There is no randomness, only insufficient resolution of structure.
- Every action, thought, or phenomenon emits a consequence ripple that feeds back to the source or reconfigures the system.

Ripple Drift

A key experimental and conceptual term. Ripple drift refers to the non-linear, field-sensitive propagation pattern of structured emissions, especially under asymmetric or constrained conditions (e.g., slits, field pressure zones).

- It explains why wave–particle duality is an illusion.
- Predicts ripple asymmetry in field zones, replacing the need for probabilistic "collapse".
- Core to future UED experimentation.

Field Alignment vs Misalignment

The central behavioral metric of all ripple systems.

- Aligned systems emit coherent ripples, resonate constructively with the field, and sustain their structure.

- Misaligned systems emit chaotic or entropic ripples, receive disruptive feedback, and eventually dissolve. This alignment principle applies across:
- Biology (homeostasis)
- Psychology (mental clarity vs emotional distortion)
- Physics (orbital stability vs decay)

Observer–Ripple Loop

A revolutionary redefinition of the observer's role. In UED, the observer is not passive, but an emitter-participant who actively shapes and is shaped by the field.

- Observation is not measurement — it is ripple engagement.
- The loop describes how the observer emits, the field responds, and the observer recalibrates, forming a cycle of awareness, action, and consequence.

Dissolution, Emission, Feedback

The three-fold life cycle of all ripple-based structures:

1. Emission – The initiation of ripple activity from any coherent source.
2. Feedback – The return signal from the field, indicating alignment or correction need.
3. Dissolution – The fading of misaligned or exhausted structures back into field neutrality.

These dynamics form the basis of UED explanations for death, memory, decay, evolution, learning, and structural persistence.

Closing Note for Appendix A

This glossary is not just a vocabulary set. It is the semantic foundation of a new way to view the universe — one where nothing is static, nothing is isolated, and everything is a ripple of intelligent consequence.

Appendix B: UED vs Major Scientific Concepts

This quick reference table distills the conceptual differences between conventional science and Universal Energy Dynamics (UED). It does not deny the utility of current models but offers a first-principles correction to their framing. This table is intended to guide researchers, educators, and advanced learners in realigning interpretation, experimentation, and language across domains.

Comparative Table: Conceptual Shifts from Science to UED

Scientific Concept	Mainstream Science Interpretation	UED Realignment
Force	An external push or pull that causes acceleration (Newtonian foundation). Treated as fundamental.	Ripple Consequence — No fundamental forces exist; motion arises from ripple tension and alignment in the SVE. Force is a perceptual artifact of energetic imbalance.
Time	Treated as a fourth dimension in spacetime (Einstein); flows uniformly in Newtonian model; dilates in relativistic speeds.	Ripple Delay — Time is not a dimension but an emergent delay in ripple feedback loops within SVE. It varies with alignment, not geometry.
Gravity	Explained as curvature of spacetime caused by mass-energy (General Relativity); affects all objects equally.	Ripple Tension Field — Mass creates localized tension in SVE, causing ripple convergence. Gravitational behavior is ripple rebalancing, not spacetime warping.
Photon	A discrete, massless particle of light carrying quantized energy ($E = hf$); dual wave–particle nature.	Resonant Ripple — No particle needed. Light is a ripple emission through SVE, observed as "photon" only at points of resonance interaction.
Probability	Core to quantum mechanics; reality exists in probabilistic superpositions until collapsed by measurement.	Consequence Patterning — No indeterminacy. Every emission has structure; what appears random is ripple misalignment or insufficient resolution of structure.
Mind and Brain	Mind is emergent from neural complexity; brain is the seat of consciousness; thoughts = neural activity.	Mind Emission Field — Mind is a structured ripple emission that modulates and is modulated by the brain. The brain is an interface, not the source.
Memory	Stored as synaptic patterns or encoded neural connections.	Ripple Imprint — Memory is retained as field resonance; recall is a harmonic realignment, not digital retrieval.
Consciousness	A mystery; often reduced to brain function or dismissed as epiphenomenon.	Ripple Awareness — Consciousness is a self-aware emission node within SVE, capable of interpreting its own consequence.
Observation	Causes wavefunction collapse in quantum mechanics; observer is often paradoxical or excluded.	Ripple Participation — Observer emits ripples and receives feedback. Observation is not collapse — it's consequence detection.
Light Travel	Light travels through space at constant speed (c); described as both wave and particle.	Ripple Alignment Drift — Light is ripple propagation; speed is field coherence limit, not universal constant. No "travel" — only propagation through SVE alignment.
Evolution	Driven by random mutation and natural selection over time.	Ripple Filtration — Evolution is the field's natural filtration of aligned vs. misaligned patterns. Sustained coherence survives.
Death	Biologically defined endpoint; irreversible loss of brain activity.	Ripple Dissolution — Loss of structural alignment; ripple pattern ceases emission and remerges into SVE neutrality. No metaphysics needed.

Interpretation Note

This table is not intended to negate current models where they are useful — but to recontextualize them as approximate constructs within a deeper field-based reality.

UED operates from first principles, not postulated particles, spacetime metaphors, or statistical abstractions. It invites a full-spectrum rethinking of not just physics, but how we define existence, observation, and consequence itself.

Appendix C: UED-Aligned Redefinitions of Common Terms

From abstraction to alignment — reinterpreting meaning through ripple logic

This appendix presents how UED redefines familiar terms not by discarding them, but by realigning their meaning with consequence-driven structure. These redefinitions remove conceptual ambiguity and unify language across science, philosophy, and experiential inquiry.

Glossary of UED-Aligned Terms

Truth → Consequence Alignment

"Truth is not what matches belief. It is what sustains alignment."
In UED, truth is not relative, dogmatic, or dependent on perception. It is the degree to which a ripple structure sustains coherence with the field.

- A belief is true only if it yields sustainable consequence
- Truth is measurable through ripple alignment, not majority opinion or historical status

Reality → Structured Field Interaction

"Reality is not a thing. It is a process of participation."

Reality is not a static environment. It is the totality of structured interactions between ripple systems within the SVE field.

- Matter, mind, memory, and motion are expressions of participation
- Reality is relational, not absolute — but it is also not arbitrary
- What exists is what persists through alignment

Intelligence → Ripple Awareness + Adaptation

"Intelligence is not knowledge. It is coherence in action."
UED defines intelligence as a system's capacity to:

1. Sense ripple feedback, and
2. Adapt emissions to maintain alignment

- A cell adapting to heat = intelligence
- A person rethinking beliefs after consequence = intelligence
- Intelligence is ripple literacy, not IQ

Emotion → Energetic Feedback Signal

"Emotions are not feelings. They are ripple diagnostics."
Emotions are real-time field signals indicating:

- Resonance (joy, peace, love)
- Interference (anger, grief, anxiety) They are not irrational. They are feedback indicators of your current alignment with energetic consequence.

Evolution → Ripple Sustainability

"Survival of the aligned, not the fittest."
Evolution is not driven by random mutation or competition. It is the field's filtration process, selecting structures that can:

- Emit stable ripple patterns
- Sustain feedback coherence
- Avoid self-disruption

Nature doesn't reward strength. It sustains balance.

Self → Temporary Ripple Structure

"You are not the content of your thoughts. You are the pattern of your participation."
In UED, the self is not fixed identity or soul. It is a temporarily stable ripple arrangement capable of interacting with the field:

- It can dissolve, reconfigure, or grow
- Ego is ripple over-identification
- Liberation = cessation of misaligned ripple emission

Mind → Emission Interface

"The mind is not a storage unit. It is a structured ripple engine."
Mind is the layered, field-aware emission mechanism of ripple identity.

- It receives consequence, translates it into awareness, and emits aligned or distorted responses
- Thoughts are not private objects — they are field-active emissions

Death → Dissolution of Ripple Identity

"What ends is not the self. What ends is emission."
Death is not annihilation or transformation. It is the ceasing of ripple emission from a coherent structure.

- No afterlife, no void — only reabsorption into SVE neutrality
- Memories may persist in field traces, but identity no longer

exists as an emitter

God → Consequence Intelligence Within the Field

"God is not a being. God is alignment itself."
UED does not define or deny God. It interprets the idea as:

- The intelligent responsiveness of the structured field
- Not a judge or creator, but a participatory intelligence embedded in consequence
- "Divinity" becomes structural awareness within SVE

Closing Insight

Words shape perception. Perception shapes participation. UED reclaims language not to redefine reality, but to reveal it — through consequence, coherence, and field responsiveness.
This glossary is not just semantic. It is philosophical correction through structural precision.

Appendix D: UED Application Matrix

Where ripple meets relevance — practical cross-domain integrations of UED

UED is not confined to abstract theory or speculative metaphysics. It is a unified energetic framework that applies directly across all domains of inquiry and practice — from particle physics to education, from brain science to ethical systems.

This appendix provides a high-level overview of how UED reframes, enriches, and unifies diverse fields of knowledge through its core principles: ripple participation, structured consequence, and observer-field interaction.

Physics

Conventional Focus	UED Application
Force, particles, spacetime, relativity	Replaces force with ripple tension, redefines particles as stable ripple constructs, reframes gravity as field imbalance, and time as ripple delay
Outcome	First-principles consistency with direct consequence logic; no dependence on invisible forces or abstract dimensions

Psychology

Conventional Focus	UED Application
Emotion, trauma, cognition, therapy	Emotions are field feedback signals, trauma is ripple entanglement, thought is emission, healing is realignment
Outcome	Moves from symptom management to energetic diagnosis and correction; makes psychology field-based and precision-guided

Neuroscience

Conventional Focus	UED Application
Brain structures, neurons, memory encoding, consciousness	Brain is a ripple modulator, not the source of mind; memory is field imprint; consciousness is observer–ripple loop
Outcome	Mind-body coherence, explains insight, intuition, and creativity through field sensitivity, not computation

Biology

Conventional Focus	UED Application
Genes, adaptation, metabolism, cellular activity	Cells are ripple-sensitive entities, DNA is memory of field-aligned patterning, evolution is sustainability of ripple coherence
Outcome	Explains non-local intelligence in plants, fungi, and organisms; shifts biology from chemistry to consequence mechanics

Cosmology

Conventional Focus	UED Application
Big Bang, expansion, dark matter, black holes	Universe as structured ripple field, galaxies as dynamic ripple loops, black holes as ripple collapse nodes, redshift as frequency stretching
Outcome	Removes need for singularities and dark energy; replaces speculative cosmology with ripple-driven evolution of structure

Ethics

Conventional Focus	UED Application
Moral rules, social norms, religious values	Ethics becomes consequence coherence: aligned actions persist, misaligned actions collapse. No dogma — only feedback.
Outcome	Provides universal, non-moralistic ethics grounded in ripple sustainability and system health

AI & Technology

Conventional Focus	UED Application
Algorithms, logic systems, machine learning	AI reinterpreted as ripple simulation engines; consciousness modeling becomes field participation, not computation
Outcome	Guides development of aligned, non-destructive AI; introduces emotional intelligence as resonance feedback, not programmed empathy

Education & Human Development

Conventional Focus	UED Application
Curriculum, intelligence testing, behavior management	Intelligence is ripple responsiveness, learning is ripple realignment, memory is resonant imprinting
Outcome	Rebuilds education systems around awareness, coherence, and ripple literacy; redefines growth as field participation skill

Closing Perspective

UED doesn't just unify knowledge. It restructures how knowledge behaves.

This matrix shows that UED is not about adding a new theory to every field — it is about replacing fragmentation with ripple fluency.

Where others see disciplines, UED sees domains of ripple participation — governed by the same field, responding to the same law:

What sustains, stays. What distorts, dissolves.

Appendix E: Proposed Experiments

Ripple logic in the lab — validating UED through structured consequence

This appendix presents brief overviews of key experimental proposals emerging from the UED framework. Each experiment is designed to test a specific principle of ripple participation, field responsiveness, or consequence alignment, offering measurable outcomes to distinguish UED logic from conventional interpretations.

These protocols emphasize simplicity, reproducibility, and openness, making them suitable for both formal research and citizen science initiatives.

1. Ripple Drift Phenomenon

Purpose: To replace wave–particle duality with observable ripple field drift patterns through asymmetric or dynamically variable slit geometries.

Setup:

- Modified double-slit apparatus
- Variable-width and shape slits (e.g., curved, tapered, or angular)
- Coherent ripple emitter (e.g., laser, water surface waves, or sound)
- High-resolution detection screen or digital field-mapping tool

Prediction (UED):

- Asymmetric interference patterns due to field-sensitive ripple engagement
- Drift and skew not explainable by photon or particle models
- Ripple behavior changes with geometry — not just with measurement or presence of an observer

Distinction: Unlike quantum models requiring wavefunction collapse, UED attributes these outcomes to field topology and ripple alignment response — not observation.

2. Structured Vacuum Energy (SVE) Field Zone Interaction

Purpose: To explore how mass or energy density alters ripple behavior in localized zones — without invoking gravity or force.
Setup:

- Test platform with suspended field-sensitive medium (vibration plate, fluid chamber, or electromagnetic resonance zone)
- Introduction of structured objects (e.g., geometric blocks, coils, harmonic materials) into neutral field
- Tracking system for ripple propagation and decay

Prediction (UED):

- Ripple velocity, coherence, and decay rate change around dense or tuned masses
- Ripple paths curve or drift due to SVE tension, not because of gravitational pull
- Certain material arrangements may amplify or dampen ripple response

Distinction: Supports the idea that mass and charge are structured field expressions, not independent properties — and that field zones can be mapped without force-based models.

3. Resonance Interference Mapping

Purpose: To chart the energetic coherence or disruption of ripple fields in real environments, spaces, or bodies.
Setup:

- Ripple emitter (sound wave, electromagnetic pulse, or harmonic signal)
- Grid of sensors (microphones, vibration detectors, EM field readers)
- Controlled introduction of sound, speech, movement, or emotional tone
- Comparative measurement in "coherent" vs "distorted" spaces (e.g., nature vs artificial interiors)

Prediction (UED):

- High-alignment environments (e.g., forests, still rooms, sacred geometry spaces) show lower interference and smoother decay
- Discordant environments show chaotic feedback and ripple fragmentation
- Human vocal tones or emotional expression modulate ripple quality detectably

Distinction: Replaces vague "vibe" or "energy" talk with measurable ripple coherence metrics.

4. Emotional Ripple Emission Study

Purpose: To validate that emotions emit ripple patterns which interact with field-sensitive systems — not just internal chemistry.
Setup:

- Participants undergo emotional induction (e.g., memory recall, music, images)
- Ripple-responsive medium (e.g., water, sound field, plasma, plant response sensors) nearby
- Non-contact, real-time ripple detection (e.g., cymatics, high-speed wave capture)

Prediction (UED):

- Emotional states cause observable ripple shifts in surrounding field
- Calmer states = harmonic patterns
- Agitated states = ripple fragmentation or resonance interference

Distinction: Provides physical evidence that emotions are not internal abstractions, but field-participating emissions.

5. Ripple Delay as Time Simulation (Perception Lab)

Purpose: To test UED's claim that time is not a fixed flow, but a ripple delay based on field alignment.
Setup:

- Visual/auditory stimulus sequence under different energetic conditions (stress, calm, coherence-enhancing environment)
- Time perception tests (interval estimation, synchronization tasks)

- Optional biofeedback to monitor internal alignment (HRV, EEG)

Prediction (UED):

- Time perception varies with ripple coherence state of observer
- Aligned observers experience compressed, flowing time
- Misaligned states result in fragmented, distorted, or elongated time perception

Distinction: Redefines time distortion not as neurological error but as consequence of field misalignment.

Closing Note

These experiments are not meant to "prove" UED in isolation — they are designed to demonstrate observable effects that standard models either cannot explain or require artificial constructs to justify.

UED doesn't promise answers. It proposes consequence-based questions that reality must respond to — and invites the world to measure, ripple by ripple.

References & Further Reading

Primary UED Works

These foundational materials present the core of Universal Energy Dynamics:

1. **The Law of Consequences (LOC)**

DOI: 10.5281/zenodo.15048775 A foundational document defining the energetic basis of consequence in structured fields.

1. **Universal Framework of Realization & Perception (UED)**

DOI: 10.5281/zenodo.15105060 A comprehensive overview of UED logic, field behavior, ripple mechanics, and observer participation.

1. **Chavan–Einstein Unified Theory (CEUT)**

DOI: 10.5281/zenodo.14992289 Integrates field logic with a restructured interpretation of relativity and quantum behavior.

1. **Structured Energy Equilibrium Principle (SEEP)**

DOI: 10.5281/zenodo.14916491 A precise articulation of how ripple equilibrium defines structural coherence across scales.

1. **Black Hole Formation – A Structured Field Model**

DOI: 10.5281/zenodo.14788539 Replaces singularity-based models with ripple convergence logic within SVE.

1. **CMPED: Planetary Energy Dynamics & Resonance Alignment**

DOI: 10.5281/zenodo.15164927 Field-level dynamics explaining orbital stability, resonance, and gravitational behavior.

Complementary Works by the Author

1. *Reality Realigned: Correcting Science, Philosophy, and Perception through UED*

A comprehensive treatise aligning UED with both ancient wisdom and modern physics.

1. *The War of Thoughts* (Series)

Features intellectual debates with historical thinkers, reinterpreting science and metaphysics through UED.

1. *Energy in Action: The Soul Demystified*

Explores mind, soul, and consciousness through ripple structuring.

1. *Manifestation: From Nothing to Everything*

A narrative exploration of existence built on first principles, rooted in the UED paradigm.

Relevant Conventional References

To understand what UED responds to and reinterprets:

- **Einstein, A.** *Relativity: The Special and the General Theory*
- **Planck, M.** *The Theory of Heat Radiation*
- **Bohr, N.**, and **Heisenberg, W.**: Foundational texts on quantum indeterminacy
- **Richard Feynman** – *QED: The Strange Theory of Light and Matter*
- **Stephen Hawking** – *A Brief History of Time*
- **Rupert Sheldrake** – *The Science Delusion* (alternative view on scientific dogma)

Suggested Further Reading & Exploration

- **David Bohm** – *Wholeness and the Implicate Order* (Non-locality and holistic physics — early precursors to field-participation models)
- **Iain McGilchrist** – *The Master and His Emissary* (Brain hemispheres and consciousness — helps bridge psychology with field behavior)
- **Thomas Kuhn** – *The Structure of Scientific Revolutions* (Understanding scientific paradigms, resistance to change,

and realignment cycles)

- **Fritjof Capra** – *The Tao of Physics* (Early attempt to bridge Eastern wisdom with quantum thinking)

Acknowledgments

This work is the result of not just thought — but consequence. It stands upon the silent ripple of every experience, every observation, and every question that refused to be ignored.

I extend my deepest gratitude to:

- My late father, Shri Jagannath Chavan, whose life of quiet alignment taught me that truth is not spoken loudly — it is lived sincerely. His values ripple through every layer of this work.
- The independent thinkers, students, and educators who believed in the possibility of a deeper science — one rooted not in dogma or prestige, but in participation, perception, and clarity.
- The scientific pioneers of the past, including Einstein, Newton, Planck, and Feynman — not as flawless icons, but as emitters of essential ripples that shaped the space we now realign.
- The ancient philosophers, mystics, and seers — across traditions — whose intuitive realizations whispered truths that formal science often bypassed.
- All who challenged me — directly or silently — forcing deeper alignment through resistance, criticism, or absence. You were necessary.

And finally, to those who will carry this forward:

You are not followers of UED. You are ripple initiators in a field that has always listened. May your work, your questions, and your presence continue the alignment.

This report is not an end. It is a beginning — of what happens when realization meets structure, and structure responds with consequence.

With respect and resonance,

Sandeep

About the Author

Sandeep is an independent researcher, educator, and author whose work spans the intersection of physics, consciousness, and existential philosophy. With a background in engineering and years of hands-on teaching experience, he has guided thousands of students through complex scientific ideas — not just to learn, but to question.

Driven by the conviction that **truth is a matter of consequence, not consensus**, Sandeep developed the **Universal Energy Dynamics (UED)** framework — a first-principles, ripple-based reinterpretation of reality that challenges both ancient mysticism and modern scientific orthodoxy. His writings and research aim to restore the observer, unify knowledge, and realign science with its original purpose: understanding through participation.

Unaffiliated with academic institutions, unconstrained by reward systems, and uninterested in prestige, Sandeep continues his work from a place of clarity, resilience, and unshakable belief in independent inquiry.

"You don't need to fight the system. You just need to emit a ripple strong enough to make the system realign."

ORCID Profile: https://orcid.org/0009-0005-4190-7164
Website: https://www.sandeepjchavan.com/

Public Use Statement

This work is created with the intent to liberate knowledge — not to own it.
The content, ideas, frameworks, and terminology within this report on Universal Energy Dynamics (UED) are made available for public use, open research, educational sharing, and collaborative development.

You are free to:

- Quote, cite, and reference this material in your work, lectures, videos, or research.
- Translate the content into other languages for educational purposes.
- Use diagrams, tables, and glossary terms in your classrooms or workshops.
- Build upon UED principles for independent exploration or experimental work.

You may not:

- Repackage or sell this content under a different authorship without explicit acknowledgment.
- Misrepresent or distort UED concepts to promote dogma, superstition, or commercial exploitation.
- Use the material to mislead or claim institutional endorsement where none exists.

Attribution

When sharing or adapting this material, please credit the author: Sandeep J Chavan, Independent Researcher – Universal Energy Dynamics (UED)

Our Position

This work belongs to everyone who is aligned with truth and inquiry. It is part of a growing movement to return science to its participatory, ethical, and consequence-rooted foundation. No permission is needed to use it for the right reasons.
The universe doesn't patent gravity. You don't need permission to ripple.

Disclaimer

This report presents the independent theoretical and philosophical framework known as Universal Energy Dynamics (UED). The content reflects the author's original research, observations, and interpretations, and is not affiliated with any academic institution, government body, or scientific authority.

Intellectual Responsibility

The ideas herein are meant to challenge, reinterpret, and expand upon conventional scientific, psychological, and philosophical models. They are offered for educational, conceptual, and exploratory purposes and do not claim to replace established science where it remains practically valid.

While every effort has been made to maintain logical coherence, clarity, and consequence-alignment, readers are encouraged to test, question, and interpret these ideas independently. The author does not promote belief, but participation and engagement.

Experimental Suggestions

All experimental outlines in this report are conceptual frameworks, not finalized laboratory procedures. Any practical implementation must follow appropriate safety protocols, ethical guidelines, and technical refinement suited to the testing environment.

Medical, Psychological, or Technological Claims

This report does not offer medical advice, diagnostic tools, or therapeutic substitutes. References to neuroscience, emotion, or psychology are theoretical constructs within the UED field model and are not intended as clinical assertions.

Liability

The author disclaims any liability for misuse, misinterpretation, or misrepresentation of the concepts presented. Any individual or organization acting on the content of this report does so at their own discretion and responsibility.

UED is a framework for re-seeing, not a system for blind acceptance. Use it with clarity, apply it with ethics, and explore it with consequence.

Don't miss out!

Visit the website below and you can sign up to receive emails whenever Sandeep Chavan publishes a new book. There's no charge and no obligation.

https://books2read.com/r/B-A-EPGPC-CFFHG

BOOKS 2 READ

Connecting independent readers to independent writers.

Did you love *The Great Realignment*? Then you should read *The Original Intent*[1] by Sandeep Chavan!

[2]

What if the world's most revered spiritual texts and philosophies were never meant to become religions?

In *The Original Intent*, independent thinker and philosophical engineer **Sandeep Chavan** takes readers on a bold, intelligent journey across time, culture, and consciousness to decode the *true purpose* behind humanity's oldest teachings. This is not a book of faith—it is a book of *clarity*. A dialogue with the past to rescue the wisdom that was buried beneath institutions, rituals, and dogmas.

Drawing from figures like **Krishna, Buddha, Vyasa, Confucius, Patanjali, Jesus, Muhammad, Socrates, Ambedkar, Vivekananda**, and many others, Chavan creates a powerful narrative arena where

1. https://books2read.com/u/b5GwMw

2. https://books2read.com/u/b5GwMw

forgotten voices return—not to preach, but to converse. Through these deeply human, intellectually fierce exchanges, the book reveals how ancient teachings were originally designed as **adaptive manuals**, not theological mandates. They were intended to guide seekers, not manufacture believers.

Each chapter tackles a vital question:

Were these teachings ever meant to form religions?Why do we need God?What is our true purpose?Who or what actually drives our actions—nature, soul, or free will?How did sacred ideas become institutionalized—and at what cost?

At the heart of the book is **Chavan's own voice**—bridging Eastern metaphysics, Western philosophy, and the cutting-edge lens of his framework **Universal Energy Dynamics (UED)**. With sharp reasoning and deep empathy, he reframes ideas like karma, dharma, the soul, intelligence, and truth as living principles rooted in consequence and awareness—not in belief.

Part of the groundbreaking *The War of Thoughts* series, *The Original Intent* is the fourth volume and stands as a pivotal work: grounding the cosmic, decoding the sacred, and rebuilding the bridge between **wisdom and relevance**.

For anyone who has ever asked, "What were our ancestors really trying to tell us?"—this book is your invitation to rediscover the answer, not in blind reverence, but in conscious reflection.

Sandeep Chavan doesn't offer conclusions. He reawakens questions that were never meant to be forgotten.

Read more at https://www.sandeepjchavan.com/.

Also by Sandeep Chavan

Chavan's UED research
CMGED Research Report
Chavan's Cosmic Continuum (C^3)
Frequency
Ripples of Reality
Reality Realigned
The Great Realignment
I – The Cosmic Leverage

Echoes of War: The Russia-Ukraine Conflict and Its Global Ripples
Shattered Lives
The War Economy
Borders Redrawn
Arms and Allies
Collateral Damage
Justice in the Ashes

Somewhere in the Darkness of the Cosmos
The Blueprint of the Universe

The Cosmic Breath

The Eternal Sage Chronicles
The Sage and the Sovereign
The Family Pillar
The People's Flame
The Warrior's Soul
The Bridging Mentor
The Guiding Light

The Exodus
The Last Cycle: 2075

The Fundamental Theory of Life & Existence
Energy in Action: The Soul Demystified
The Active Mind
Energy in Connection
Death: The Disconnection
Subconsciousness: The Chosen Disconnection
Desires: The Engine of Life

The Restoration of Human Knowledge
Restoring Bhagavad Gita

The War of Thoughts

Cracks in the Known
The Human Game
The Divine Dilemma
The Original Intent
The Answers

Ultra Modern Life Philosophy
The Duality of Laughter and Tears
Take It Easy
Compare to Accept

Standalone
It's Not AI, It's AHI - Amplified Human Intelligence
The Decision Paradox
The Four Sapiens
Malicious Script of Indian Polity
The IIT Legacy & Global Impact
The IIT Legacy & Global Impact
You are just a Version of Your Original
Win the Game You Didn't Choose
Always Keep Your Bags Packed
Reveal to Shield - Learn the Game of Facades
Engage Beyond Elections
Culture, Identity & Change: The Evolution of Indian Society
The Triangular Dynamics
Never Trust Without Doubt
If You Can't Change the Country, Leave the Country
Live Fearless with Fear
India: The Investment Magnet
Realignment

Powering a Billion Dreams
Universal Skill
Extension: From Brush to Algorithm
Human Side of Einstein
The Future Knows No Lines
Let's Bring the Generations to the Table
Unified Energy Dynamics
A Beat That Never Dies
Strategic Horizons
Right From Scratch
The Forgotten Innovators
Physics Fixing
The Savior Illusion
The Selfish Side
The Age of the Weakest Sapiens
Neural Wars
The Absorption: There Is No Death
Strange Therapist
Bubbles: Become a Drop
Beyond the Blue
Half-Scholar
The Law of Truth
Failure
Philosophy 2.0: A Structural Manifesto
Chaos: A Gentle Guide to the Unresolvable
Self-Help: New Syllabus

www.ingramcontent.com/pod-product-compliance
Lightning Source LLC
LaVergne TN
LVHW090401210926
848480LV00002B/412